Parallel Programming and Concurrency with C# 10 and .NET 6

A modern approach to building faster, more responsive, and asynchronous .NET applications using C#

Alvin Ashcraft

BIRMINGHAM—MUMBAI

Parallel Programming and Concurrency with C# 10 and .NET 6

Associate Group Product Manager: Gebin George
Publishing Product Manager: Sathyanarayanan Ellapulli
Senior Editor: Ruvika Rao
Content Development Editor: Yashi Gupta
Technical Editor: Maran Fernandes
Copy Editor: Safis Editing
Project Coordinator: Manisha Singh
Proofreader: Safis Editing
Indexer: Hemangini Bari
Production Designer: Roshan Kawale
Marketing Coordinator: Sonakshi Bubbar

First published: August 2022
Production reference: 1120822

Published by Packt Publishing Ltd.
Livery Place
35 Livery Street
Birmingham
B3 2PB, UK.

ISBN 978-1-80324-367-2
www.packt.com

To my wife, Stelene, for supporting me through the process of writing a second book and for helping me become my best self in our journey together. To my three daughters, for working harder and smarter than I did at their ages and for their patience during my writing process.

– Alvin Ashcraft

Contributors

About the author

Alvin Ashcraft is a writer, software engineer, and developer community champion with over 27 years of experience in software and content development. He has worked with Microsoft Windows, web, and cloud technologies since 1995 and has been awarded as a Microsoft MVP 13 times.

Alvin works remotely in the Philadelphia area for Microsoft as a content developer on Microsoft Learn for the Windows developer documentation team. He also helps organize the TechBash developer conference in the Northeast US. He has previously worked for companies such as Oracle, Genzeon, and Allscripts. Originally from the Allentown, PA, area, Alvin currently resides in West Grove, PA, with his wife and daughters.

I want to thank the people who have been close to me and supported me,
especially my parents and my wife, Stelene's, parents.

About the reviewers

Ricardo Peres is a Portuguese developer, blogger, and book author and is currently a team leader at Dixons Carphone. He has over 20 years of experience in software development and his interests include distributed systems, architectures, design patterns, and .NET development. He won the Microsoft MVP award in 2015 and held this title up to 2020.

He also authored *Entity Framework Core Cookbook – Second Edition* and *Mastering ASP.NET Core 2.0* and was a technical reviewer for *Learning NHibernate 4* for Packt. He also contributed to Syncfusion's Succinctly collection, with titles on .NET development. Ricardo maintains a blog—*Development With A Dot*—where he writes about technical issues. You can keep up with him on Twitter at @rjperes75.

Joseph Guadagno is a senior director of engineering at Rocket Mortgage, the US's largest mortgage lender, based in Detroit, Michigan. He has been writing software for over 30 years and has been an active member of the .NET community, serving as a Microsoft MVP in .NET, for more than 10 years. He has spoken throughout the United States and at international events on topics including Microsoft .NET, Microsoft Azure, and SQL. You can see the complete list of events he has spoken at at https://www.josephguadagno.net/presentations.

When not sitting at a computer, Joe loves to hang out with his family and play games. You can connect with Joe on Twitter at @jguadagno, on Facebook at JosephGuadagnoNet, and on his blog at https://www.josephguadagno.net.

Table of Contents

3

Best Practices for Managed Threading

4

User Interface Responsiveness and Threading

Part 2: Parallel Programming and Concurrency with C#

5

Asynchronous Programming with C#

6

Parallel Programming Concepts

7

Task Parallel Library (TPL) and Dataflow

8

Parallel Data Structures and Parallel LINQ

9

Working with Concurrent Collections in .NET

Part 3: Advanced Concurrency Concepts

10

Debugging Multithreaded Applications with Visual Studio

11

Canceling Asynchronous Work

12

Unit Testing Async, Concurrent, and Parallel Code

Assessments

Index

Other Books You May Enjoy

Preface

Parallel programming and concurrency have become prevalent in modern software development. In this book, you will learn how to leverage the latest asynchronous, parallel, and concurrency features in .NET 6 when building your next application. We will explore the power of multithreaded C# development patterns and practices. By exploring the benefits and challenges of threading in .NET through concise, real-world examples, choosing the right option for your project will become second nature.

You have many choices when introducing multithreading to a new or existing .NET application. The goal of this book is to not only teach you how to use parallel programming and concurrency in C# and .NET but also to help you understand which of the constructs to choose for a given scenario. Whether you are developing for desktop, mobile, the web, or the cloud, performance and responsiveness are key to the success of an application. This book will help every type of C# developer to scale their applications to their users' needs and avoid the pitfalls often encountered with multithreaded development.

Who this book is for

This book is for beginner- to intermediate-level .NET developers who want to employ the latest parallel and concurrency features in .NET when building their applications. You should have a solid understanding of the C# language and some version of the .NET Framework or .NET Core.

What this book covers

Chapter 1, *Managed Threading Concepts*, covers the basics of working with managed threading in .NET. We will discuss how to create and destroy threads, handle exceptions, synchronize data, and the objects provided by .NET to handle background operations. You will gain a basic understanding of how threads can be managed in a .NET application. Practical examples in this chapter will illustrate how to use managed threading in C# projects.

Chapter 2, *Evolution of Multithreaded Programming in .NET*, introduces some of the concepts and features that will be explored in more depth in later chapters, including async/await, concurrent collections, and parallelism. You will learn how their options are expanded when selecting how to approach concurrency in applications.

Chapter 3, *Best Practices for Managed Threading*, covers some best practices when it comes to integrating managed threading concepts. We will cover important concepts such as static data, deadlocks, and exhausting managed resources. These are all areas that can lead to unstable applications and unexpected behavior. You will be given practical advice to avoid these pitfalls.

Chapter 4, User Interface Responsiveness with Threading, explains how to use ThreadPool in .NET. The real-world examples in this chapter will give you valuable options for ensuring UI responsiveness in your .NET applications.

Chapter 5, Asynchronous Programming with C#, explains asynchronous programming in C# and explores the best use of tasks in .NET.

Chapter 6, Parallel Programming Concepts, delves deeper into the **Task Parallel Library** (**TPL**) and tasking concepts.

Chapter 7, Task Parallel Library (TPL) and Dataflow, introduces the TPL Dataflow Library and illustrates some common patterns for its use through in-depth examples.

Chapter 8, Parallel Data Structures and Parallel LINQ (PLINQ), explores some of .NET's useful features, including **Parallel LINQ** (**PLINQ**). Follow along with some practical examples of PLINQ in C#.

Chapter 9, Working with Concurrent Collections in .NET, dives deeper into some of the concurrent collections that help provide data integrity when using concurrency and parallelism in your code.

Chapter 10, Debugging Multithreaded Applications with Visual Studio, teaches you how to use the power of Visual Studio when debugging multithreaded .NET applications. This chapter will explore the tools in detail through concrete examples.

Chapter 11, Canceling Asynchronous Work, dives deeper into the different methods available to cancel concurrent and parallel work with .NET. You will gain a deep understanding of how to safely cancel asynchronous work.

Chapter 12, Unit Testing Async, Concurrent, and Parallel Code, provides some concrete advice and real-world examples of how developers can unit test code that employs multithreaded constructs. These examples will illustrate how unit tests can still be reliable while covering code that performs multithreaded operations.

To get the most out of this book

To follow along with the examples in this book, the following software is recommended for Windows developers:

- Visual Studio 2022 version 17.0 or later
- .NET 6

While these are recommended, if you have the .NET 6 SDK installed, you can use your preferred editor for most of the examples. For example, Visual Studio 2022 for Mac on macOS 10.13 or later, JetBrains Rider, or Visual Studio Code will work just as well. However, for any WPF or WinForms projects, Visual Studio and Windows are required. Newer versions of Visual Studio and .NET, when they are released, should also work with the examples in this book.

You are expected to have a foundational knowledge of C# and .NET with a working knowledge of **Language Integrated Query (LINQ)**.

Software/hardware covered in the book	Operating system requirements
Visual Studio 2022 (version 17.0 or later)	Windows 10/11
Visual Studio Code	Windows, macOS, or Linux
.NET 6	Windows, macOS, or Linux

The most recent Visual Studio 2022 install instructions and prerequisites can always be found on Microsoft Docs here: https://docs.microsoft.com/visualstudio/install/install-visual-studio?view=vs-2022.

If you are using the digital version of this book, we advise you to type the code yourself or access the code from the book's GitHub repository (a link is available in the next section). Doing so will help you avoid any potential errors related to the copying and pasting of code.

If you are unfamiliar with LINQ, there is a great C# reference on Microsoft Docs to get you started before working through the examples in this book: https://docs.microsoft.com/dotnet/csharp/programming-guide/concepts/linq/.

After reading this book, I would also recommend exploring the posts on the .NET Parallel Programming team blog. Most of the articles are several years old, but they explore the thinking behind many of the decisions made when building the .NET libraries that expose parallel programming constructs: https://devblogs.microsoft.com/pfxteam/.

Download the example code files

You can download the example code files for this book from GitHub at https://github.com/PacktPublishing/Parallel-Programming-and-Concurrency-with-C-sharp-10-and-.NET-6. If there's an update to the code, it will be updated in the GitHub repository.

We also have other code bundles from our rich catalog of books and videos available at https://github.com/PacktPublishing/. Check them out!

Download the color images

We also provide a PDF file that has color images of the screenshots and diagrams used in this book. You can download it here: https://packt.link/Z4GcQ.

Conventions used

There are a number of text conventions used throughout this book.

`Code in text`: Indicates code words in the text, database table names, folder names, filenames, file extensions, pathnames, dummy URLs, user input, and Twitter handles. Here is an example: "By calling `ThreadPool.SetMaxThreads`, you can change the maximum values for `workerThreads` and `completionPortThreads`."

A block of code is set as follows:

```
public async Task PerformCalculations()
{
    _runningTotal = 3;
    await MultiplyValue().ContinueWith(async (Task) => {
        await AddValue();
        });
    Console.WriteLine($"Running total is {_runningTotal}");
}
```

When we wish to draw your attention to a particular part of a code block, the relevant lines or items are set in bold:

```
private async Task MultiplyValue()
{
    await Task.Delay(100);
    var currentTotal = Interlocked.Read(ref
        _runningTotal);
    Interlocked.Exchange(ref _runningTotal,
        currentTotal * 10);
}
}
```

Any command-line input or output is written as follows:

```
$ mkdir css
$ cd css
```

Bold: Indicates a new term, an important word, or words that you see onscreen. For instance, words in menus or dialog boxes appear in **bold**. Here is an example: "Let's look at a quick example of how to implement this in our **CancellationPatterns** project."

> **Tips or important notes**
> Appear like this.

Get in touch

Feedback from our readers is always welcome.

General feedback: If you have questions about any aspect of this book, email us at customercare@ packtpub.com and mention the book title in the subject of your message.

Errata: Although we have taken every care to ensure the accuracy of our content, mistakes do happen. If you have found a mistake in this book, we would be grateful if you would report this to us. Please visit www.packtpub.com/support/errata and fill in the form.

Piracy: If you come across any illegal copies of our works in any form on the internet, we would be grateful if you would provide us with the location address or website name. Please contact us at copyright@packt.com with a link to the material.

If you are interested in becoming an author: If there is a topic that you have expertise in and you are interested in either writing or contributing to a book, please visit authors.packtpub.com.

Share Your Thoughts

Once you've read *Parallel Programming and Concurrency with C# 10 and .NET 6*, we'd love to hear your thoughts! Scan the QR code below to go straight to the Amazon review page for this book and share your feedback.

https://packt.link/r/1803243678

Your review is important to us and the tech community and will help us make sure we're delivering excellent quality content.

Part 1: Introduction to Threading in .NET

In this part, you will learn the basics of managed threading in .NET, discover how it has evolved since the early days of the .NET Framework, and pick up some best practices to avoid common pitfalls.

This part contains the following chapters:

- *Chapter 1, Managed Threading Concepts*
- *Chapter 2, Evolution of Multithreaded Programming in .NET*
- *Chapter 3, Best Practices for Managed Threading*
- *Chapter 4, User Interface Responsiveness with Threading*

1

Managed Threading Concepts

Parallel programming and **concurrency** are becoming more prevalent in modern .NET development. Most developers today have been exposed to **asynchronous programming** with the `async` and `await` keywords in C#. This book will cover all of these concepts in the chapters ahead.

In this chapter, we will start with the basics of how to work with **managed threading** in .NET. You will learn how to create and destroy threads, handle exceptions, synchronize data, and utilize the objects provided by .NET to handle background operations. Additionally, you will gain a basic understanding of how threads can be managed in a .NET application. The practical examples in this chapter will illustrate how to leverage managed threading in C# projects.

In this chapter, we will cover the following topics:

- .NET threading basics
- Creating and destroying threads
- Handling threading exceptions
- Synchronizing data across threads
- Scheduling and canceling work

By starting with the core concepts of threading in .NET, you will gain a solid foundation as you move forward with your learning throughout this book. It is important to understand the basics to prevent common mistakes from being made while introducing threading and asynchrony to .NET applications. It is all too easy to exhaust resources or put the application's data in an invalid state. Let's get started with managed threading with C#.

Technical requirements

To follow along with the examples in this chapter, the following software is recommended:

- Visual Studio 2022 version 17.0 or later

- .NET 6

While these are recommended, as long as you have .NET 6 installed, you can use your preferred editor. For example, Visual Studio 2022 for Mac, JetBrains Rider, or Visual Studio Code will work just as well.

All the code examples for this chapter can be found on GitHub at `https://github.com/PacktPublishing/Parallel-Programming-and-Concurrency-with-C-sharp-10-and-.NET-6/tree/main/chapter01`.

.NET threading basics

It's time to get started by learning about the basics of threading in C# and .NET. We will be covering the managed threading concepts that are available in .NET 6, but many of these features have been part of .NET since the beginning. The `System.Threading` namespace has been available since .NET Framework 1.0. In the subsequent 20 years, there have been many useful features added for developers.

In order to responsibly use threading in your applications, you should understand exactly what a **thread** is and how threads are used by your application's **processes**.

Threads and processes

We will start our journey with the basic units of application processing, threads, and processes. A process encapsulates all the execution of an application. This is true for all platforms and frameworks. In .NET, you can think of a process as your `.exe` or hosted service.

> **Note**
>
> In .NET Framework, the concept of **application domains** (or app domains), which create isolation units within a process, was introduced. These app domains provide security and reliability by isolating the execution of code loaded into a new app domain. App domains still exist but are not available for developers to create or unload in .NET Core and modern versions of .NET. To read more about app domains, check out this Microsoft Docs article at `https://docs.microsoft.com/dotnet/framework/app-domains/application-domains`.

A **thread** represents a single unit of execution within a process. By default, a .NET application will execute all its logic on a single thread (that is, the primary or main thread). Developers can leverage managed threads and other .NET constructs to move from a single-threaded to a multithreaded world, but how do you know when to take this step?

When should we use multithreading in .NET?

There are multiple factors to consider when deciding whether to introduce threading to an application. These factors are both internal and external to the application. The external factors include the hardware in terms of where the application will be deployed, how powerful the processors are where the application will be running, and what other types of processes will be running on these systems?

If your application will be competing for limited resources, it is best to be judicious with the use of multiple threads. If users get the impression that your application is impacting the performance of their systems, you will need to scale back on the number of threads being consumed by your process. Another factor that comes into play is how critical your application is in relation to others on the system. Mission-critical applications will have more resources allocated to remain responsive when needed.

Other common reasons for introducing threading relate to the application itself. Desktop and mobile applications need to keep the **user interface** (**UI**) responsive to user input. If the application needs to process a large amount of data or load it from a database, file, or network resource, executing on the main thread can cause the UI to freeze or lag. Also, executing long-running tasks in parallel on multiple threads can reduce the overall execution time of the task.

These operations can be offloaded to **background threads** if the execution of the tasks is not critical to the application state. Let's look at the difference between foreground threads and background threads in .NET.

Background threads

The difference between foreground threads and background threads might not be what you think. A managed thread created as a foreground thread is not the UI thread or the main thread. Foreground threads are threads that will prevent the managed process from terminating if they are running. If an application is terminated, any running background threads will be stopped so that the process can shut down.

By default, newly created threads are foreground threads. To create a new background thread, set the `Thread.IsBackground` property to `true` before starting the thread. Additionally, you can use the `IsBackground` property to determine the background status of an existing thread. Let's look at an example where you might want to use a background thread in your application.

In this example, we will create a console application in Visual Studio that will continuously check the status of a network connection on a background thread. Create a new .NET 6 console app project, name it `BackgroundPingConsoleApp`, and in `Program.cs`, enter the following code:

```
Console.WriteLine("Hello, World!");
var bgThread = new Thread(() =>
{
    while (true)
    {
        bool isNetworkUp = System.Net.NetworkInformation
            .NetworkInterface.GetIsNetworkAvailable();
        Console.WriteLine($"Is network available? Answer:
            {isNetworkUp}");
        Thread.Sleep(100);
    }
});
bgThread.IsBackground = true;
bgThread.Start();
for (int i = 0; i < 10; i++)
{
    Console.WriteLine("Main thread working...");
    Task.Delay(500);
}
Console.WriteLine("Done");
Console.ReadKey();
```

Let's discuss each part of the preceding code before we run it and examine the output:

1. The first `Console.WriteLine` statement was created by the project template. We'll keep this here to help illustrate the order output in the console.

2. Next, we're creating a new `Thread` type named `bgThread`. Inside the body of the thread, there is a `while` loop that will execute continuously until the thread is terminated. Inside the loop, we're calling the `GetIsNetworkAvailable` method and outputting the result of that call to the console. Before starting over again, we're using `Thread.Sleep` to inject a 100-millisecond delay.

3. The next line after creating the thread is the key part of this lesson:

    ```
    bgThread.IsBackground = true;
    ```

Setting the IsBackground property to true is what makes our new thread a background thread. This tells our application that the code executing inside the thread is not critical to the application, and the process can terminate without needing to wait for the thread to complete its work. That is a good thing here because the while loop we created will never be complete.

4. On the next line, we start the thread with the Start method.

5. Next, the application kicks off some work inside the application's primary thread. A for loop will execute 10 times and output "Main thread working..." to the console. At the end of each iteration of the loop, Task.Delay is used to wait 500 milliseconds, hopefully providing some time for the background thread to perform some work.

6. After the for loop, the application will output "Done" to the console and wait for the user input to terminate the application with the Console.ReadKey method.

Now, run the application and examine the console output. You can press any key to stop the application when you feel you have let it run for long enough:

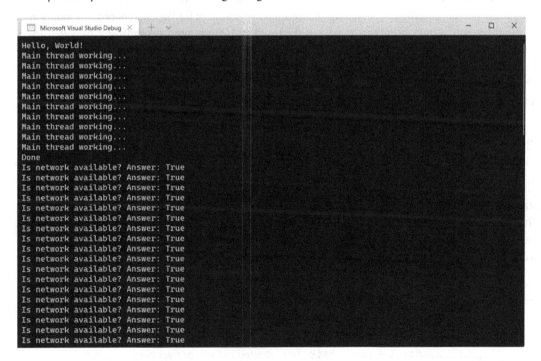

Figure 1.1 – Viewing the threaded console application output

The result might not be what you expected. You can see that the program executed all the logic on the primary thread before starting any of the background thread work. Later, we'll see how to change the priority of the threads to manipulate which work will be processed first.

What is important to understand, in this example, is that we were able to stop the console application by pressing a key to execute the `Console.ReadKey` command. Even though the background thread is still running, the process does not consider the thread to be critical to the application. If you comment out the following line, the application will no longer terminate by pressing a key:

```
bgThread.IsBackground = true;
```

The application will have to be stopped by closing the command window or using the **Debug | Stop Debugging** menu item in Visual Studio. Later, in the *Scheduling and canceling work* section, we will learn how to cancel work in a managed thread.

Before we look at any more examples of using managed threads, we will take some time to learn exactly what they are.

What is managed threading?

In .NET, **managed threading** is implemented by the `System.Threading.Thread` class that we used in the previous example. The managed execution environment for the current process monitors all the threads that have been run as part of the process. **Unmanaged threading** is how threads are managed when programming in C++ with native Win32 threading elements. It is possible for unmanaged threads to enter a managed process through **COM interop** or through platform invoke (`PInvoke`) calls from .NET code. If this thread is entering the managed environment for the first time, .NET will create a new `Thread` object to be managed by the execution environment.

A managed thread can be uniquely identified using the `ManagedThreadId` property of the `Thread` object. This property is an integer that is guaranteed to be unique across all threads and will not change over time.

The `ThreadState` property is a read-only property that provides the current execution state of the `Thread` object. In the example in the *.NET threading basics* section, if we had checked the `ThreadState` property before calling `bgThread.Start()`, it would have been `Unstarted`. After calling `Start`, the state will change to `Background`. If it were not a background thread, calling `Start` would change the `ThreadState` property to `Running`.

Here is a full list of the `ThreadState` enum values:

- `Aborted`: The thread has been aborted.
- `AbortRequested`: An abort has been requested but has not yet been completed.
- `Background`: The thread is running in the background (`IsBackground` has been set to true).
- `Running`: The thread is currently running.

- `Stopped`: The thread has been stopped.

- `StopRequested`: A stop has been requested but has not yet been completed.

- `Suspended`: The thread has been suspended.

- `SuspendRequested`: Thread suspension has been requested but has not yet been completed.

- `Unstarted`: The thread has been created but not yet started.

- `WaitSleepJoin`: The thread is currently blocked.

The `Thread.IsAlive` property is a less specific property that can tell you whether a thread is currently running. It is a `boolean` property that will return `true` if the thread has started and has not been stopped or aborted in some way.

Threads also have a `Name` property that defaults to `null` if they have never been set. Once a `Name` property is set on a thread, it cannot be changed. If you attempt to set the `Name` property of a thread that is not null, it will throw `InvalidOperationException`.

We will cover additional aspects of managed threads in the remainder of this chapter. In the next section, we will dive deeper into the available methods and options for creating and destroying threads in .NET.

Creating and destroying threads

Creating and destroying threads are fundamental concepts of managed threading in .NET. We have already seen one code example that created a thread, but there are some additional constructors of the `Thread` class that should be discussed first. Also, we will look at a few methods of pausing or interrupting thread execution. Finally, we will cover some ways to destroy or terminate a thread's execution.

Let's get started by going into more detail regarding creating and starting threads.

Creating managed threads

Creating **managed threads** in .NET is accomplished by instantiating a new `Thread` object. The `Thread` class has four constructor overloads:

- `Thread(ParameterizedThreadStart)`: This creates a new `Thread` object. It does this by passing a delegate with a constructor that takes an object as its parameter that can be passed when calling `Thread.Start()`.

- `Thread(ThreadStart)`: This creates a new `Thread` object that will execute the method to be invoked, which is provided as the `ThreadStart` property.

- `Thread(ParameterizedThreadStart, Int32)`: This adds a `maxStackSize` parameter. Avoid using this overload because it is best to allow .NET to manage the stack size.

- Thread(ThreadStart, Int32): This adds a maxStackSize parameter. Avoid using this overload because it is best to allow .NET to manage the stack size.

Our first example used the Thread(ThreadStart) constructor. Let's look at a version of that code that uses ParameterizedThreadStart to pass a value by limiting the number of iterations of the while loop:

```
Console.WriteLine("Hello, World!");
var bgThread = new Thread((object? data) =>
{
    if (data is null) return;
    int counter = 0;
    var result = int.TryParse(data.ToString(),
        out int maxCount);
    if (!result) return;
    while (counter < maxCount)
    {
        bool isNetworkUp = System.Net.NetworkInformation
            .NetworkInterface.GetIsNetworkAvailable();
        Console.WriteLine($"Is network available? Answer:
            {isNetworkUp}");
        Thread.Sleep(100);
        counter++;
    }
});
bgThread.IsBackground = true;
bgThread.Start(12);
for (int i = 0; i < 10; i++)
{
    Console.WriteLine("Main thread working...");
    Task.Delay(500);
}
Console.WriteLine("Done");
Console.ReadKey();
```

If you run the application, it will run just like the last example, but the background thread should only output 12 lines to the console. You can try passing different integer values into the Start method to see how that impacts the console output.

If you want to get a reference to the thread that is executing the current code, you can use the `Thread.CurrentThread` static property:

```
var currentThread = System.Threading.Thread.CurrentThread;
```

This can be useful if your code needs to check the current thread's `ManagedThreadId`, `Priority`, or whether it is running in the background.

Next, let's look at how we can pause or interrupt the execution of a thread.

Pausing thread execution

Sometimes, it is necessary to pause the execution of a thread. A common real-life example of this is a retry mechanism on a background thread. If you have a method that sends log data to a network resource, but the network is unavailable, you can call `Thread.Sleep` to wait for a specific interval before trying again. `Thread.Sleep` is a static method that will block the current thread for the number of milliseconds specified. It is not possible to call `Thread.Sleep` on a thread other than the current one.

We have already used `Thread.Sleep` in the examples in this chapter, but let's change the code slightly to see how it can impact the order of events. Change the `Thread.Sleep` interval inside the thread to `10`, remove the code that makes it a background thread, and change the `Task.Delay()` call to `Thread.Sleep(100)`:

```
Console.WriteLine("Hello, World!");
var bgThread = new Thread((object? data) =>
{
    if (data is null) return;
    int counter = 0;
    var result = int.TryParse(data.ToString(), out int
        maxCount);
    if (!result) return;
    while (counter < maxCount)
    {
        bool isNetworkUp = System.Net.NetworkInformation.
            NetworkInterface.GetIsNetworkAvailable();
        Console.WriteLine($"Is network available? Answer:
            {isNetworkUp}");
        Thread.Sleep(10);
        counter++;
```

```
        }
    });
    bgThread.Start(12);
    for (int i = 0; i < 12; i++)
    {
        Console.WriteLine("Main thread working...");
        Thread.Sleep(100);
    }
    Console.WriteLine("Done");
    Console.ReadKey();
```

When running the application again, you can see that putting a greater delay on the primary thread allows the process inside bgThread to begin executing before the primary thread completes its work:

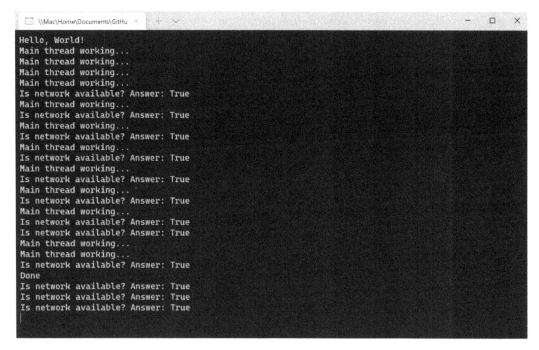

Figure 1.2 – Using Thread.Sleep to change the order of events

The two Thread.Sleep intervals can be adjusted to see how they impact the console output. Give it a try!

Additionally, it is possible to pass `Timeout.Infinite` to `Thread.Sleep`. This will cause the thread to pause until it is interrupted or aborted by another thread or the managed environment. Interrupting a blocked or paused thread is accomplished by calling `Thread.Interrupt`. When a thread is interrupted, it will receive a `ThreadInterruptedException` exception.

The exception handler should allow the thread to continue working or clean up any remaining work. If the exception is unhandled, the runtime will catch the exception and stop the thread. Calling `Thread.Interrupt` on a running thread will have no effect until that thread has been blocked.

Now that you understand how to create an interrupt thread, let's wrap up this section by learning how to destroy a thread.

Destroying managed threads

Generally, **destroying a managed thread** is considered an unsafe practice. That is why .NET 6 no longer supports the `Thread.Abort` method. In .NET Framework, calling `Thread.Abort` on a thread would raise a `ThreadAbortedException` exception and stop the thread from running. Aborting threads was not made available in .NET Core or any of the newer versions of .NET. If some code needs to be forcibly stopped, it is recommended that you run it in a separate process from your other code and use `Process.Kill` to terminate the other process.

Any other thread termination should be handled cooperatively using cancelation. We will see how to do this in the *Scheduling and canceling work* section. Next, let's discuss some of the exceptions to handle when working with managed threads.

Handling threading exceptions

There are a couple of exception types that are specific to managed threading, including the `ThreadInterruptedException` exception that we covered in the previous section. Another exception type that is specific to threading is `ThreadAbortException`. However, as we discussed in the previous section, `Thread.Abort` is not supported in .NET 6, so, although this exception type exists in .NET 6, it is not necessary to handle it, as this type of exception is only possible in .NET Framework applications.

Two other exceptions are the `ThreadStartException` exception and the `ThreadStateException` exception. The `ThreadStartException` exception is thrown if there is a problem starting the managed thread before any user code in the thread can be executed. The `ThreadStateException` exception is thrown when a method on the thread is called that is not available when the thread is in its current `ThreadState` property. For example, calling `Thread.Start` on a thread that has already started is invalid and will cause a `ThreadStateException` exception. These types of exceptions can usually be avoided by checking the `ThreadState` property before acting on the thread.

It is important to implement comprehensive exception handling in multithreaded applications. If code in managed threads begins to fail silently without any logging or causing the process to terminate, the application can fall into an invalid state. This can also result in degrading performance and unresponsiveness. While this kind of degradation might be noticed quickly for many applications, some services, and other non-GUI-based applications, could continue for some time without any issues being noticed. Adding logging to the exception handlers along with a process to alert users when logs are reporting failures will help to prevent problems with undetected failing threads.

In the next section, we'll discuss another challenge with multithreaded code: keeping data in-sync across multiple threads.

Synchronizing data across threads

In this section, we will look at some of the methods that are available in .NET for synchronizing data across multiple threads. Shared data across threads can be one of the primary pain points of multithreaded development if not handled properly. Classes in .NET that have protections in place for threading are said to be **thread-safe**.

Data in multithreaded applications can be synchronized in several different ways:

- **Synchronized code regions**: Only synchronize the block of code that is necessary using the `Monitor` class or with some help from the .NET compiler.

- **Manual synchronization**: There are several **synchronization primitives** in .NET that can be used to manually synchronize data.

- **Synchronized context**: This is only available in .NET Framework and Xamarin applications.

- **System.Collections.Concurrent classes**: There are specialized .NET collections to handle concurrency. We will examine these in *Chapter 9*.

In this section, we'll look at the first two methods. Let's start by discussing how to synchronize code regions in your application.

Synchronizing code regions

There are several techniques you can use to synchronize regions of your code. The first one we will discuss is the `Monitor` class. You can surround a block of code that can be accessed by multiple threads with calls to `Monitor.Enter` and `Monitor.Exit`:

```
...
Monitor.Enter(order);
order.AddDetails(orderDetail);
```

```
Monitor.Exit(order);

...
```

In this example, imagine you have an `order` object that is being updated by multiple threads in parallel. The `Monitor` class will lock access from other threads while the current thread adds an `orderDetail` item to the `order` object. The key to minimizing the chance of introducing wait time to other threads is by only locking the lines of code that need to be synchronized.

> **Note**
>
> The `Interlocked` class, as discussed in this section, performs atomic operations in user mode rather than kernel mode. If you want to read more about this distinction, I recommend checking out this blog post by Nguyen Thai Duong: `https://duongnt.com/interlocked-synchronization/`.

The `Interlocked` class provides several methods for performing atomic operations on objects shared across multiple threads. The following list of methods is part of the `Interlocked` class:

- `Add`: This adds two integers, replacing the first one with the sum of the two
- `And`: This is a bitwise `and` operation for two integers
- `CompareExchange`: This compares two objects for equality and replaces the first if they are equal
- `Decrement`: This decrements an integer
- `Exchange`: This sets a variable to a new value
- `Increment`: This increments an integer
- `Or`: This is a bitwise `or` operation for two integers

These `Interlocked` operations will lock access to the target object only for the duration of that operation.

Additionally, the `lock` statement in C# can be used to lock access to a block of code to only a single thread. The `lock` statement is a language construct implemented using the .NET `Monitor.Enter` and `Monitor.Exit` operations.

There is some built-in compiler support for the `lock` and `Monitor` blocks. If an exception is thrown inside one of these blocks, the lock is automatically released. The C# compiler generates a `try/finally` block around the synchronized code and makes a call to `Monitor.Exit` in the `finally` block.

Let's finish up this section on synchronization by looking at some other .NET classes that provide support for manual data synchronization.

Manual synchronization

The use of **manual synchronization** is common when synchronizing data across multiple threads. Some types of data cannot be protected in other ways, such as these:

- **Global fields**: These are variables that can be accessed globally across the application.

- **Static fields**: These are static variables in a class.

- **Instance fields**: These are instance variables in a class.

These fields do not have method bodies, so there is no way to put a synchronized code region around them. With manual synchronization, you can protect all the areas where these objects are used. These regions can be protected with `lock` statements in C#, but some other synchronization primitives provide access to shared data and can coordinate the interactions between threads on a more granular level. The first construct we will examine is the `System.Threading.Mutex` class.

The `Mutex` class is similar to the `Monitor` class in that it blocks access to a region of code, but it can also provide the ability to grant access to other processes. When using the `Mutex` class, use the `WaitOne()` and `ReleaseMutex()` methods to acquire and release the lock. Let's look at the same order/order details example. This time, we'll use a `Mutex` class declared at the class level:

```
private static Mutex orderMutex = new Mutex();
...
orderMutex.WaitOne();
order.AddDetails(orderDetail);
orderMutex.ReleaseMutex();
...
```

If you want to enforce a timeout period on the `Mutex` class, you can call the `WaitOne` overload with a timeout value:

```
orderMutex.WaitOne(500);
```

It is important to note that `Mutex` is a **disposable** type. You should always call `Dispose()` on the object when you are finished using it. Additionally, you can also enclose a disposable type within a `using` block to have it disposed of indirectly.

In this section, the last .NET manual locking construct we are going to examine is the `ReaderWriterLockSlim` class. You can use this type if you have an object that is used across multiple threads, but most of the code is reading data from the object. You don't want to lock access to the object in the blocks of code that are reading data, but you do want to prevent reading while the object is being updated or simultaneously written. This is referred to as "multiple readers, single writer."

This ContactListManager class contains a list of contacts that can be added to or retrieved by a phone number. The class assumes that these operations can be called from multiple threads and uses the ReaderWriterLockSlim class to apply a read lock in the GetContactByPhoneNumber method and a write lock in the AddContact method. The locks are released in a finally block to ensure they are always released, even when exceptions are encountered:

```
public class ContactListManager
{
    private readonly List<Contact> contacts;
    private readonly ReaderWriterLockSlim contactLock =
        new ReaderWriterLockSlim();
    public ContactListManager(
        List<Contact> initialContacts)
    {
        contacts = initialContacts;
    }
    public void AddContact(Contact newContact)
    {
        try
        {
            contactLock.EnterWriteLock();
            contacts.Add(newContact);
        }
        finally
        {
            contactLock.ExitWriteLock();
        }
    }
    public Contact GetContactByPhoneNumber(string
        phoneNumber)
    {
        try
        {
            contactLock.EnterReadLock();
            return contacts.FirstOrDefault(x =>
                x.PhoneNumber == phoneNumber);
        }
```

```
        finally
        {
            contactLock.ExitReadLock();
        }
    }
}
```

If you were to add a `DeleteContact` method to the `ContactListManager` class, you would leverage the same `EnterWriteLock` method to prevent any conflicts with the other operations in the class. If a lock is forgotten in one usage of `contacts`, it can cause any of the other operations to fail. Additionally, it is possible to apply a timeout to the `ReaderWriterLockSlim` locks:

```
contacts.EnterWriteLock(1000);
```

There are several other synchronization primitives that we have not covered in this section, but we have discussed some of the most common types that you will use. To read more about the available types for manual synchronization, you can visit Microsoft Docs at `https://docs.microsoft.com/dotnet/standard/threading/overview-of-synchronization-primitives`.

Now that we have examined different ways of synchronizing data when working with managed threads, let's cover two more important topics before wrapping up this first chapter. We are going to discuss techniques to schedule work on threads and how to cancel managed threads cooperatively.

Scheduling and canceling work

When orchestrating multithreaded processing in an application, it is important to understand how to schedule and cancel work on managed threads.

Let's start by looking at how scheduling works with managed threads in .NET.

Scheduling managed threads

When it comes to managed threads, scheduling is not as explicit as it might sound. There is no mechanism to tell the operating system to kick off work at specific times or to execute within certain intervals. While you could write this kind of logic, it is probably not necessary. The process of scheduling managed threads is simply managed by setting priorities on the threads. To do this, set the `Thread.Priority` property to one of the available `ThreadPriority` values: `Highest`, `AboveNormal`, `Normal` (default), `BelowNormal`, or `Lowest`.

Generally, higher priority threads will execute before those of lower priority. Usually, a thread of `Lowest` priority will not execute until all the higher priority threads have been completed. If the `Lowest` priority thread has started and a `Normal` thread kicks off, the `Lowest` priority thread will be suspended so that the `Normal` thread can be run. These rules are not absolute, but you can use them as a guide. Most of the time, you will leave the default of `Normal` for your threads.

When there are multiple threads of the same priority, the operating system will cycle through them, giving each thread up to a maximum allotment of time before suspending work and moving on to the next thread of the same priority. The logic will vary by the operating system, and the prioritization of a process can change based on whether the application is in the foreground of the UI.

Let's use our network checking code to test thread priorities:

1. Start by creating a new console application in Visual Studio

2. Add a new class to the project, named `NetworkingWork`, and add a method named `CheckNetworkStatus` with the following implementation:

```
public void CheckNetworkStatus(object data)
{
    for (int i = 0; i < 12; i++)
    {
        bool isNetworkUp = System.Net.
            NetworkInformation.NetworkInterface
                .GetIsNetworkAvailable();
        Console.WriteLine($"Thread priority
            {(string)data}; Is network available?
                Answer: {isNetworkUp}");
        i++;
    }
}
```

The calling code will be passing a parameter with the priority of the thread that is currently executing the message. That will be added as part of the console output inside the `for` loop, so users can see which priority threads are running first.

3. Next, replace the contents of `Program.cs` with the following code:

```
using BackgroundPingConsoleApp_sched;
Console.WriteLine("Hello, World!");
var networkingWork = new NetworkingWork();
```

```
var bgThread1 = new
    Thread(networkingWork.CheckNetworkStatus);
var bgThread2 = new
    Thread(networkingWork.CheckNetworkStatus);
var bgThread3 = new
    Thread(networkingWork.CheckNetworkStatus);
var bgThread4 = new
    Thread(networkingWork.CheckNetworkStatus);
var bgThread5 = new
    Thread(networkingWork.CheckNetworkStatus);
bgThread1.Priority = ThreadPriority.Lowest;
bgThread2.Priority = ThreadPriority.BelowNormal;
bgThread3.Priority = ThreadPriority.Normal;
bgThread4.Priority = ThreadPriority.AboveNormal;
bgThread5.Priority = ThreadPriority.Highest;
bgThread1.Start("Lowest");
bgThread2.Start("BelowNormal");
bgThread3.Start("Normal");
bgThread4.Start("AboveNormal");
bgThread5.Start("Highest");
for (int i = 0; i < 10; i++)
{
    Console.WriteLine("Main thread working...");
}
Console.WriteLine("Done");
Console.ReadKey();
```

The code creates five Thread objects, each with a different Thread.Priority value. To make things a little more interesting, the threads are being started in reverse order of their priorities. You can try changing this on your own to see how the order of execution is impacted.

4. Now run the application and examine the output:

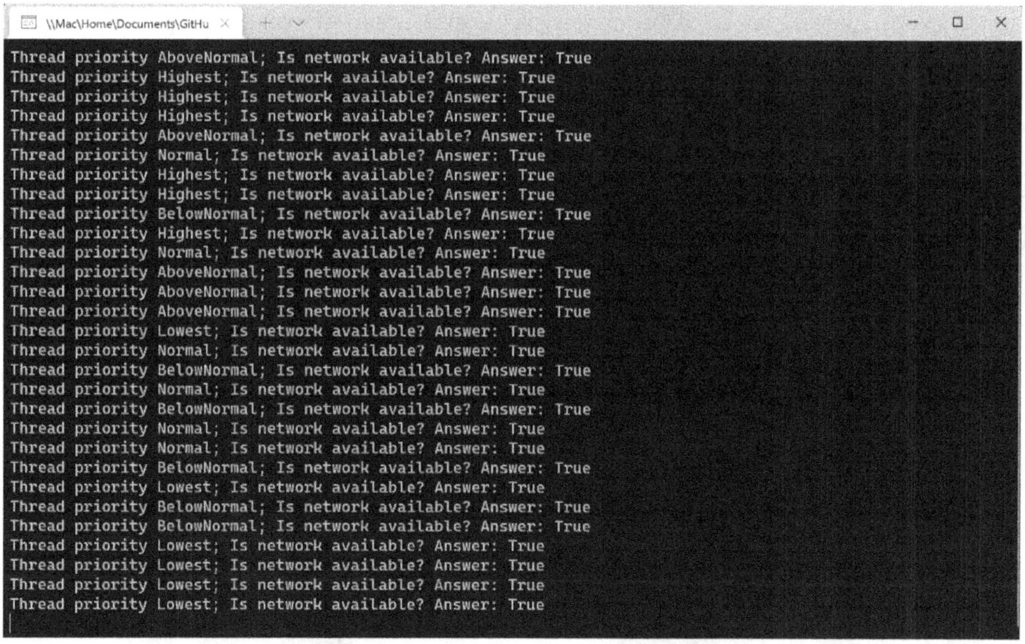

```
Thread priority AboveNormal; Is network available? Answer: True
Thread priority Highest; Is network available? Answer: True
Thread priority Highest; Is network available? Answer: True
Thread priority Highest; Is network available? Answer: True
Thread priority AboveNormal; Is network available? Answer: True
Thread priority Normal; Is network available? Answer: True
Thread priority Highest; Is network available? Answer: True
Thread priority Highest; Is network available? Answer: True
Thread priority BelowNormal; Is network available? Answer: True
Thread priority Highest; Is network available? Answer: True
Thread priority Normal; Is network available? Answer: True
Thread priority AboveNormal; Is network available? Answer: True
Thread priority AboveNormal; Is network available? Answer: True
Thread priority AboveNormal; Is network available? Answer: True
Thread priority Lowest; Is network available? Answer: True
Thread priority Normal; Is network available? Answer: True
Thread priority BelowNormal; Is network available? Answer: True
Thread priority Normal; Is network available? Answer: True
Thread priority BelowNormal; Is network available? Answer: True
Thread priority Normal; Is network available? Answer: True
Thread priority Normal; Is network available? Answer: True
Thread priority BelowNormal; Is network available? Answer: True
Thread priority Lowest; Is network available? Answer: True
Thread priority BelowNormal; Is network available? Answer: True
Thread priority BelowNormal; Is network available? Answer: True
Thread priority Lowest; Is network available? Answer: True
Thread priority Lowest; Is network available? Answer: True
Thread priority Lowest; Is network available? Answer: True
Thread priority Lowest; Is network available? Answer: True
```

Figure 1.3 – Console output from five different threads

You can see that the operating system, which, in my case, is Windows 11, sometimes executes lower priority threads before all the higher priority threads have completed their work. The algorithm for selecting the next thread to run is a bit of a mystery. You should also remember that this is multithreading. Multiple threads are running at once. The exact number of threads that can run simultaneously will vary by the processor or virtual machine configuration.

Let's wrap things up by learning how to cancel a running thread.

Canceling managed threads

Canceling managed threads is one of the more important concepts to understand about managed threading. If you have long-running operations running on foreground threads, they should support cancelation. There are times when you might want to allow users to cancel the processes through your application's UI, or the cancelation might be part of a cleanup process while the application is closing.

To cancel an operation in a managed thread, you will use a `CancellationToken` parameter. The `Thread` object itself does not have built-in support for cancellation tokens like some of the modern threading constructs .NET. So, we will have to pass the token to the method running in the newly created thread. In the next exercise, we will modify the previous example to support cancelation:

1. Start by updating `NetworkingWork.cs` so that the parameter passed to `CheckNetworkStatus` is a `CancellationToken` parameter:

    ```
    public void CheckNetworkStatus(object data)
    {
        var cancelToken = (CancellationToken)data;
        while (!cancelToken.IsCancellationRequested)
        {
            bool isNetworkUp = System.Net
                .NetworkInformation.NetworkInterface
                    .GetIsNetworkAvailable();
            Console.WriteLine($"Is network available?
                Answer: {isNetworkUp}");
        }
    }
    ```

    The code will keep checking the network status inside a while loop until `IsCancellationRequested` becomes `true`.

2. In `Program.cs`, we're going to return to working with only one `Thread` object. Remove or comment out all of the previous background threads. To pass the `CancellationToken` parameter to the `Thread.Start` method, create a new `CancellationTokenSource` object, and name it `ctSource`. The cancellation token is available in the `Token` property:

    ```
    var pingThread = new
        Thread(networkingWork.CheckNetworkStatus);
    var ctSource = new CancellationTokenSource();
    pingThread.Start(ctSource.Token);

    . . .
    ```

3. Next, inside the `for` loop, add a `Thread.Sleep(100)` statement to allow `pingThread` to execute while the main thread is suspended:

```
for (int i = 0; i < 10; i++)
{
    Console.WriteLine("Main thread working...");
    Thread.Sleep(100);
}
```

4. After the `for` loop is complete, invoke the `Cancel()` method, join the thread back to the main thread, and dispose of the `ctSource` object. The `Join` method will block the current thread and wait for `pingThread` to complete using this thread:

```
...
ctSource.Cancel();
pingThread.Join();
ctSource.Dispose();
```

5. Now, when you run the application, you will see the network checking stops shortly after the final `Thread.Sleep` statement on the main thread has been executed:

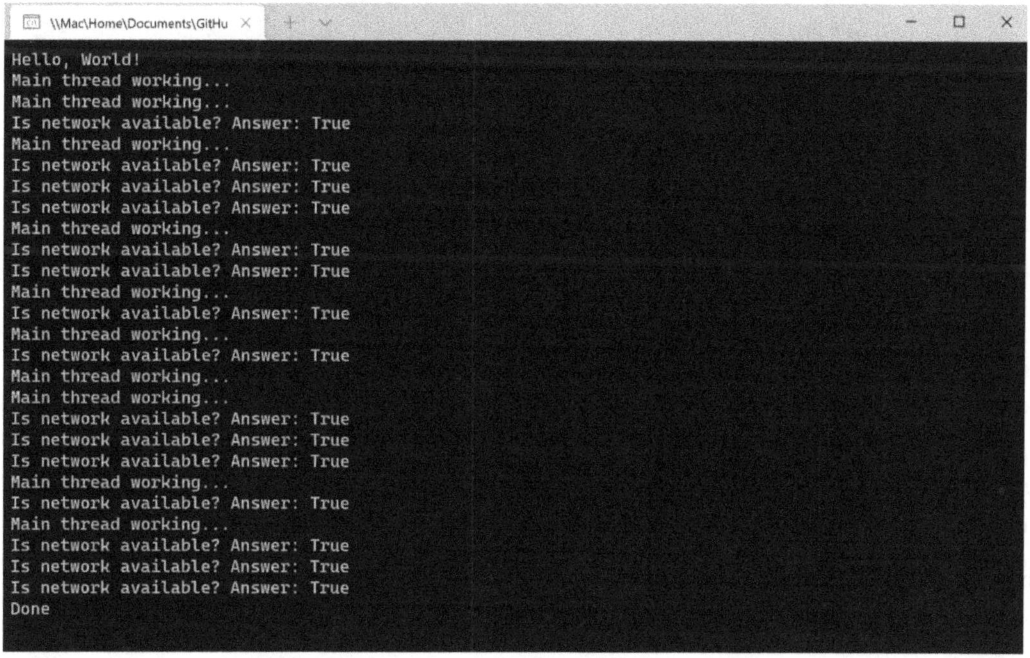

Figure 1.4 – Canceling a thread in the console application

Now the network checker application is gracefully canceling the threaded work before listening for a keystroke to close the application.

When you have a long-running process on a managed thread, you should check for cancellation as the code iterates through loops, begins a new step in a process, and at other logical checkpoints in the process. If the operation uses a timer to periodically perform work, the token should be checked each time the timer executes.

Another way to listen for cancellation is by registering a **delegate** to be invoked when a cancellation has been requested. Pass the delegate to the Token.Register method inside the managed thread to receive a cancellation callback. The following CheckNetworkStatus2 method will work exactly like the previous example:

```
public void CheckNetworkStatus2(object data)
{
    bool finish = false;
    var cancelToken = (CancellationToken)data;
    cancelToken.Register(() => {
        // Clean up and end pending work
        finish = true;
    });
    while (!finish)
    {
        bool isNetworkUp = System.Net.NetworkInformation
            .NetworkInterface.GetIsNetworkAvailable();
        Console.WriteLine($"Is network available? Answer:
            {isNetworkUp}");
    }
}
```

Using a delegate like this is more useful if you have multiple parts of your code that need to listen for a cancellation request. A **callback** method can call several cleanup methods or set another flag that is monitored throughout the thread. It encapsulates the cleanup operation nicely.

We will revisit cancellation in *Chapter 11*, as we introduce new parallelism and concurrency concepts. However, this section should provide a solid foundation for understanding what comes next.

That concludes the final section on managed threads. Let's wrap things up and review what we have learned.

Summary

In this chapter, we covered the basics of managed threading and the `System.Threading.Thread` class. You should now have a good understanding of how to create and schedule a thread in .NET. You learned about some of the techniques for passing data to threads and how to use background threads for non-critical operations, so they don't prevent your application from terminating. Finally, we used two different techniques for canceling threads in .NET.

In the next chapter, *Chapter 2*, we will learn how .NET has simplified and improved parallel programming and concurrency for developers over the last 20 years. Significant improvements were added in .NET 4.5 in the form of the `async` and `await` keywords, and .NET Core removed some of the legacy threading constructs of .NET Framework.

Questions

1. What is a managed thread?
2. How do you make a background thread?
3. What happens if you try to set the `IsBackground` property of a running thread?
4. How does .NET handle scheduling managed threads?
5. What is the highest thread priority?
6. What happens to a thread when you call `Thread.Abort()` in .NET 6?
7. How can you pass data to a method in a new thread?
8. How do you register a callback to be invoked when a cancellation is requested on a thread?

2

Evolution of Multithreaded Programming in .NET

As .NET and C# have evolved over the last 20 years, new and innovative approaches to multithreaded programming have been introduced. C# has added new language features to support asynchronous programming, and .NET Framework and .NET Core have added new types to support the languages. The most impactful improvements were introduced with C# 5 and .NET Framework 4.0 when Microsoft added the **Task Parallel Library (TPL)**, thread-safe collections, and the `async` and `await` keywords.

This chapter will introduce concepts and features that will be explored in greater depth in subsequent chapters. These concepts include the .NET **thread pool**, asynchronous programming with `async` and `await`, concurrent collections, and parallelism. We will start by discovering when and why threading features were added to .NET and C#. Then, we will create some practical examples of how to use the new concepts. Finally, we will wrap up the chapter by discussing when it makes sense to use these new features in your own projects. It is important to select the best tool for every real-world scenario.

In this chapter, you will learn about the following:

- .NET threading through the years
- Beyond threading basics
- Introduction to parallelism
- Introduction to concurrency
- Basics of `async` and `await`
- Choosing the right path forward

By the end of this chapter, you will have learned how your options have expanded when selecting how to approach concurrency in your .NET applications.

Technical requirements

To follow along with the examples in this chapter, the following software is recommended for Windows users:

- Visual Studio 2022 version 17.0 or later.

- .NET 6.

- To use the WorkingWithTimers project, you will need to install the Visual Studio workload for **.NET desktop development**.

While these are recommended, if you have .NET 6 installed, you can use your preferred editor. For example, Visual Studio 2022 for Mac on macOS 10.13 or later, JetBrains Rider, or Visual Studio Code will work just as well.

All the code examples for this chapter can be found on GitHub at `https://github.com/ PacktPublishing/Parallel-Programming-and-Concurrency-with-C- sharp-10-and-.NET-6/tree/main/chapter02`.

Let's start the chapter with a .NET and C# history lesson.

.NET threading through the years

Working with threads in .NET and C# has undergone much evolution since .NET Framework 1.0 and C# 1.0 were introduced in 2002. Most of the concepts discussed in *Chapter 1*, regarding the `System.Threading.Thread` objects have been available since those early days of .NET. While the `Thread` object is still available in .NET 6 and can be useful for simple scenarios, there are more elegant and modern solutions that are available today.

This section will highlight when the most impactful parallelism and concurrency features were added. We will begin by skipping ahead 8 years to 2010.

C# 4 and .NET Framework 4.0

In 2010, Microsoft released Visual Studio 2010 alongside C# 4 and .NET Framework 4.0. While some earlier language and framework features such as **generics**, **lambda expressions**, and **anonymous methods** would help facilitate later threading features, these 2010 releases were the most significant for threading since 2002. .NET Framework included the following features that will be explored in more detail in the subsequent sections and chapters:

- **Thread-safe collections**: This collection was added to the `System.Collections. Concurrent` namespace to provide safe access to collections of data in multithreaded code.

- **Parallel class**: This provided support for parallel loops via `Parallel.For` and `Parallel. ForEach` and for invoking parallel operations with `Parallel.Invoke`.

- **Parallel LINQ (PLINQ)**: This exposed a parallel implementation of the LINQ operations with extensions such as `AsParallel`, `WithCancellation`, and `WithDegreeOfParallelism`.

We will cover these features in the *Introduction to concurrency* and *Introduction to parallelism* sections. Next, we will learn about the important threading features that were included in .NET and C# two years later.

C# 5 and 6 and .NET Framework 4.5.x

In 2012, Microsoft released what could be considered the most important feature for modern multithreaded programming with .NET: asynchronous programming with `async` and `await`. The `async` and `await` keywords were added to C# 5 in the same release when .NET Framework 4.5 added the TPL. The centerpiece of the TPL was the `Task` class in the new `System.Threading.Tasks` namespace.

The `Task` object returns from an `async` operation, providing a way for developers to check the status of the operation or wait for its completion. The work of an `async` task is performed on a background thread on the thread pool, rather than in the main thread. We will learn more about thread pools in the *Beyond threading basics* section. The basics of the TPL will be discussed in the *Basics of async and await* section of this chapter and in more depth in *Chapter 5*.

Some tooling and language features related to async programming were added in the following years. In 2013, .NET Framework 4.5.1 was released. This release corresponded to the release of Visual Studio 2013, which added async debugging features to the **Call Stack** and **Tasks** windows. The C# 6 and Visual Studio 2015 releases added the ability for developers to await asynchronous operations in the `catch` and `finally` blocks of exception handlers.

The next features came in 2017 with Microsoft's continued shift from .NET Framework to .NET Core.

C# 7.x and .NET Core 2.0

The second major version of .NET Core released by the .NET team included the new `ValueTask` and `ValueTask<TResult>` types. A `ValueTask` type is a structure that wraps a `Task` or an `IValueTaskSource` instance and includes some additional fields. It is only available when using C# 7.0 or later. The `ValueTask` type was added because many async operations, in practice, complete synchronously but still incur the overhead of allocating a `Task` instance to return to the caller. In these cases, performance can be improved by replacing `Task` with `ValueTask`, which does not incur any allocation when completing its work synchronously. To read more about the motivation behind the introduction of `ValueTask` and when to use it, you can read the following blog post by Stephen Toub of the .NET team: `https://devblogs.microsoft.com/dotnet/understanding-the-whys-whats-and-whens-of-valuetask/`.

> **Note**
>
> If you are not familiar with Stephen Toub, he is a Partner Software Engineer for Microsoft and works as a developer on the .NET Team. His work on the .NET team was key in bringing `async`, `await`, and the TPL to the .NET developer community. You can read some of his other articles on the .NET Parallel Programming blog at `https://devblogs.microsoft.com/pfxteam/author/toub/`.

C# 7.0 also introduced **discards** to the language. A discard in C# is represented by a single underscore character (_) to replace an intentionally unused variable. A **standalone discard** replaces the need for a declared variable to hold the `Task` instance returned by an async call. By using a discard in this scenario, it signals to the compiler explicitly that you want to ignore the returned `Task` instance. Discards can be used as placeholders for variables in other scenarios, too. Using discards can make the intent of your code clearer and, in some cases, reduce memory allocation. You can read more about their use on the Microsoft Docs website at `https://docs.microsoft.com/dotnet/csharp/fundamentals/functional/discards`.

Later in 2017, C# 7.1 was released, adding a feature of note for async programming: the ability to declare the `Main` method of a class as async. This made it possible to await other async methods directly from the `Main` method.

The next async features of note came along in 2019 with C# 8.

C# 8 and .NET Core 3.0

When C# 8 and .NET Core 3.0 were released in 2019, several languages and .NET features were added to support the new **async streams** feature. As the name implies, async streams allow developers to use the new `IAsyncEnumerable` type to provide a streaming source of asynchronous data.

Let's examine a code snippet that uses `IAsyncEnumerable`:

```
public async IAsyncEnumerable<Order>
    GetLargeOrdersForCustomerAsync(int custId)
{
    await foreach (var order in
        GetOrdersByCustomerAsync(custId))
    {
        if (order.Items.Count > 10) yield return order;
    }
}
```

In this example, the new `await foreach` language feature is used to call an async method to get all the orders for a customer. It then uses a `yield return` operation to return each `order` object with more than 10 items via the `IAsyncEnumerable` type as it is processed. We will cover some more real-world scenarios of using `IAsyncEnumerable` in *Chapter 5*.

The other async feature added in C# 8 was the `System.IAsyncDisposable` interface. When implementing `IAsyncDisposable`, your class must implement a parameterless `DisposeAsync` method. If your class consumes managed resources that implement `IAsyncDisposable`, and they cannot be disposed of in line with an `async using` block, you should implement `IAsyncDisposable` and clean up these resources in a protected `DisposeAsyncCore` method. For a comprehensive example that uses both `IDisposable` and `IAsyncDisposable`, you can review the Microsoft Docs example at `https://docs.microsoft.com/dotnet/standard/garbage-collection/implementing-disposeasync#implement-both-dispose-and-async-dispose-patterns`.

This brings us to the most recent releases of C# and .NET. Let's review what's new for async developers in these 2021 releases.

C# 10 and .NET 6

.NET 6 was released in November 2021 along with C# 10. One of the new features in .NET 6 was the ability of `System.Text.Json` to serialize and deserialize an `IAsyncEnumerable` type. Prior to .NET 6, a serialized `IAsyncEnumerable` type would contain an empty JSON object. This is considered a breaking change in .NET 6, but it is a change for the better. The primary motivation behind the change was to support `IAsyncEnumerable<T>` responses in the ASP.NET Core MVC controller methods.

The other .NET 6 feature of note for async developers is that the C# project templates in Visual Studio 2021 were modernized to leverage several recent language features, including the `async Main` method available in C# 7.1 and later. The .NET team blogged about these updated templates when .NET 6 release candidate 2 was released in October 2021: `https://devblogs.microsoft.com/dotnet/announcing-net-6-release-candidate-2/#net-sdk-c-project-templates-modernized`.

This should give you an idea of when each of the significant threading features was added to C# and .NET, and it sets the stage for the upcoming sections of this chapter, where we will cover some of the basics of parallel programming and concurrency. Let's begin by looking at some more features of threading, starting with the .NET thread pool.

Beyond threading basics

Before we introduce parallel programming, concurrency, and async programming with .NET and C#, we have a few more threading concepts to cover. The most important of these is the .NET managed thread pool, which is used by awaited method calls that execute asynchronously in C#.

Managed thread pool

The `ThreadPool` class in the `System.Threading` namespace has been part of .NET since the beginning. It provides developers with a pool of worker threads that they can leverage to perform tasks in the background. In fact, that is one of the key characteristics of thread pool threads. They are background threads that run at the default priority. When one of these threads completes its task, it is returned to the pool of available threads to await its next task. You can queue as many tasks to the thread pool as the available memory will support, but the number of active threads is limited by the number that the operating system can allocate to your application, based on the processor capacity and other running processes.

If you were to use the `ThreadPool` class in a .NET 6 application, you would typically do so through the TPL, but let's explore how it can be used directly with `ThreadPool.QueueUserWorkItem`. The following code takes the example scenario of *Chapter 1*, but uses a `ThreadPool` thread to perform the background process:

```
Console.WriteLine("Hello, World!");
ThreadPool.QueueUserWorkItem((o) =>
{
    for (int i = 0; i < 20; i++)
    {
        bool isNetworkUp = System.Net.NetworkInformation.
            NetworkInterface.GetIsNetworkAvailable();
        Console.WriteLine($"Is network available? Answer:
            {isNetworkUp}");
        Thread.Sleep(100);
    }
});
for (int i = 0; i < 10; i++)
{
    Console.WriteLine("Main thread working...");
    Task.Delay(500);
}
Console.WriteLine("Done");
Console.ReadKey();
```

Here, the key differences are that there is no need to set `IsBackground` to `true`, and you do not call `Start()`. The process will start either as soon as the item is queued on `ThreadPool` or when the next `ThreadPool` becomes available. While you might not explicitly use `ThreadPool` frequently in your code, it is leveraged by many of the common threading features in .NET. So, it's important to have some understanding of how it works.

One of the common .NET features that use `ThreadPool` is timers.

Threading and timers

In this section, we will examine two timer classes that use `ThreadPool`, `System.Timers.Timer` and `System.Threading.Timer`. Both of these types are safe to use with managed threading and are available on every platform supported by .NET 6.

> **Note**
>
> Some additional timers are only applicable to either web or Windows platform development. This section will focus on the timers that are platform agnostic. To read more about the other timers, you can refer to the documentation on the Microsoft Docs website at `https://docs.microsoft.com/dotnet/standard/threading/timers`.

System.Timers.Timer

You are probably most familiar with the `Timer` object in the `System.Timers` namespace. This timer will raise an `Elapsed` event on a thread pool thread at the interval specified in the `Interval` property. The mechanism can be stopped or started by using the Boolean `Enabled` property. If you need the `Elapsed` event to only fire once, the `AutoReset` property can be set to `false`.

> **Note**
>
> To follow along with the code in this example, download the code from the **WorkingWithTimers** project of this chapter's GitHub repository: `https://github.com/PacktPublishing/Parallel-Programming-and-Concurrency-with-C-sharp-10-and-.NET-6/tree/main/chapter02`.

This example uses a `Timer` object to check for new messages and alert a user if any are found:

1. Start by declaring a `Timer` object and setting it up in an `InitializeTimer` method:

    ```
    private System.Timers.Timer? _timer;
    private void InitializeTimer()
    {
    ```

```
            _timer = new System.Timers.Timer
            {
                Interval = 1000
            };

            _timer.Elapsed += _timer_Elapsed;
    }
```

2. Next, create the _timer_Elapsed event handler to check for messages and update the users:

```
    private void _timer_Elapsed(object? sender,
        System.Timers.ElapsedEventArgs e)
    {
        int messageCount = CheckForNewMessageCount();
        if (messageCount > 0)
        {
            AlertUser(messageCount);
        }
    }
```

3. After the _timer object has its Enabled property set to true, the Elapsed event will fire every second. In the **WorkingWithTimers** project, the state is being controlled by the StartTimer() and StopTimer() methods in the TimerSample class:

```
    public void StartTimer()
    {
        if (_timer == null)
        {
            InitializeTimer();
        }
        if (_timer != null && !_timer.Enabled)
        {
            _timer.Enabled = true;
        }
    }
    public void StopTimer()
    {
        if (_timer != null && _timer.Enabled)
```

```
        {
            _timer.Enabled = false;
        }
    }
}
```

4. Run the **WorkingWithTimers** project and try using the **Start Timer** and **Stop Timer** buttons.

    You should see messages in Visual Studio's debug **Output** window appearing every second while the timer is enabled.

Note

Remember that timer events are firing on a thread pool thread. The code executing in these methods might not have access to update the UI. These timer examples are part of a **Windows Forms (WinForms)** project. The most common way to update the UI with WinForms is by checking `InvokeRequired` on the form or user control and then updating the UI with the `Invoke` method, if necessary. More information about how to update a WinForms UI can be found on the Microsoft Docs website at `https://docs.microsoft.com/dotnet/desktop/winforms/controls/how-to-make-thread-safe-calls`.

In your own applications, you would use the `AlertUser` method to present an alert message to the user or update a notification icon in the UI. Next, let's try the `System.Threading.Timer` class.

System.Threading.Timer

Now, we will create the same example with the `System.Threading.Timer` class. This `Timer` class must be initialized a little differently:

Note

To follow along with the code in this example, download the code from the **WorkingWithTimers** project of this chapter's GitHub repository: `https://github.com/PacktPublishing/Parallel-Programming-and-Concurrency-with-C-sharp-10-and-.NET-6/tree/main/chapter02`.

---

1. Start by creating a new `InitializeTimer` method:

```
private void InitializeTimer()
{
    var updater = new MessageUpdater();
    _timer = new System.Threading.Timer(
    callback: new TimerCallback(TimerFired),
    state: updater,
```

```
    dueTime: 500,
    period: 1000);
}
```

The constructor for the `Timer` class takes four parameters. The `callback` parameter is a delegate to invoke on the thread pool when the timer period elapses. The `state` parameter is an object to pass to the callback delegate. The `dueTime` parameter tells the timer how long (in milliseconds) to wait before triggering the timer for the first time. Finally, the `period` parameter specifies the interval (in milliseconds) between each delegate invocation.

2. After instantiating the timer, it will immediately start. There is no `Enabled` property to start or stop this timer. When you are done with it, you should dispose of it with either the `Dispose` method or the `DisposeAsync` method. This is happening in our `DisposeTimerAsync` method:

```
public void StartTimer()
{
    if (_timer == null)
    {
        InitializeTimer();
    }
}
public async Task DisposeTimerAsync()
{
    if (_timer != null)
    {
        await _timer.DisposeAsync();
    }
}
```

3. `MessageUpdater` is a class that is used as the `state` object provided to the `TimerCallback` method. It has a single method that handles updates to the message count. The logic to update the user about new messages can be encapsulated by this class. In our case, it will simply update the debug output with the number of new messages:

```
internal class MessageUpdater
{
    internal void Update(int messageCount)
    {
        Debug.WriteLine($"You have {messageCount} new
```

```
                            messages!");
        }
    }
```

4. The final piece to examine is the `TimerFired` callback method:

```
private void TimerFired(object? state)
{
    int messageCount = CheckForNewMessageCount();
    if (messageCount > 0 &&
        state is MessageUpdater updater)
    {
        updater.Update(messageCount);
    }
}
```

Similar to the `_timer_Elapsed` method from the previous example, this method simply checks for new messages and triggers an update. However, this time, the update is performed by the `MessageUpdater` class, which, in your application, could be abstracted through an `IMessageUpdater` interface and injected into this class for improved separation of concerns and testability.

5. Try this example by using the **Start Threading Timer** and **Stop Threading Timer** buttons in the application. You should see a debug message appearing with new message counts in the **Output** window, as you did in the previous example.

The two timers serve similar purposes; however, most of the time, you will want to use `System.Threading.Timer` to leverage its async nature. However, if you need to frequently stop and start your timer processes, the `System.Timers.Timer` class is a better choice.

Now that we have covered some additional managed threading concepts to level-set your knowledge, it's time to shift gears and introduce the concept of parallel programming with C#.

Introduction to parallelism

While exploring the history of threading in C# and .NET, we learned that parallelism was introduced to developers in .NET Framework 4.0. In this section, the aspects that will be explored are exposed in the TPL through the `System.Threading.Tasks.Parallel` class. In addition, we will cover some of the basics of PLINQ through examples. These data parallelism concepts will be covered in greater detail with real-world examples in *Chapter 6*, *Chapter 7*, and *Chapter 8*.

At a high level, parallelism is the concept of executing multiple tasks in parallel. These tasks could be related to one another, but this is not a requirement. In fact, related tasks running in parallel run a greater risk of encountering synchronization issues or blocking one another. For example, if your application loads order data from an orders service and user preferences and application state from an **Azure blob store**, these two processes can be run in parallel without having to worry about conflicts or data synchronization. On the other hand, if the application is loading order data from two different order services and combining the results in a single collection, you will need a synchronization strategy.

That type of scenario will be discussed, in greater depth, in *Chapter 9*. In this section, we will prepare for those advanced scenarios by learning some uses of the `Parallel` class. Let's start with `Parallel.Invoke`.

Using Parallel.Invoke

`Parallel.Invoke` is a method that can execute multiple actions, and they could be executed in parallel. There is no guarantee of the order in which they will execute. Each action will be queued in the thread pool. The `Invoke` call will return when all the actions have been completed.

In this example, the `Parallel.Invoke` call will execute four actions: another method in the `ParallelInvokeExample` class named `DoComplexWork`, a lambda expression, an `Action` declared inline, and a `delegate`. Here is the complete `ParallelInvokeExample` class:

```
internal class ParallelInvokeExample
{
    internal void DoWorkInParallel()
    {
        Parallel.Invoke(
            DoComplexWork,
            () => {
                Console.WriteLine($"Hello from lambda
                expression. Thread id:
                {Thread.CurrentThread.ManagedThreadId}");
            },
            new Action(() =>
            {
                Console.WriteLine($"Hello from Action.
                Thread id: {Thread.CurrentThread
                .ManagedThreadId}");
            }),
            delegate ()
```

```
        {
            Console.WriteLine($"Hello from delegate.
            Thread id: {Thread.CurrentThread
            .ManagedThreadId}");
        }
    );
    }
    private void DoComplexWork()
    {
        Console.WriteLine($"Hello from DoComplexWork
        method. Thread id: {Thread.CurrentThread
        .ManagedThreadId}");
    }
}
```

Creating a new instance of `ParallelInvokeExample` and executing `DoWorkInParallel` from a console application will produce an output similar to the following, although the order of operations may vary:

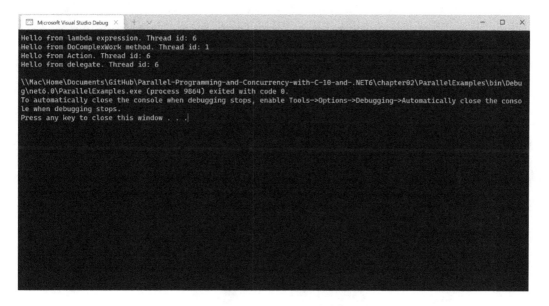

Figure 2.1 – Output produced by the DoWorkInParallel method

In the next section, we will learn how to implement a `Parallel.ForEach` loop and discuss when you might want to leverage it.

Using Parallel.ForEach

`Parallel.ForEach` is probably the most used member of the `Parallel` class in .NET. This is because, in many cases, you can simply take the body of a standard `foreach` loop and use it inside a `Parallel.ForEach` loop. However, when introducing any parallelism into a code base, you must be sure that the code being invoked is thread-safe. If the body of a `Parallel.ForEach` loop modifies any of the collections, you will either need to employ one of the synchronization methods discussed in *Chapter 1*, or use one of .NET's concurrent collections. We will introduce concurrent collections in the *Introduction to concurrency* section.

As an example of using `Parallel.ForEach`, we will create a method that accepts a list of numbers and checks whether each number is contained in a string representation of the current time:

```
internal void ExecuteParallelForEach(IList<int> numbers)
{
    Parallel.ForEach(numbers, number =>
    {
        bool timeContainsNumber = DateTime.Now.
            ToLongTimeString().Contains(number.ToString());
        if (timeContainsNumber)
        {
            Console.WriteLine($"The current time contains
            number {number}. Thread id: {Thread.
            CurrentThread.ManagedThreadId}");
        }
        else
        {
            Console.WriteLine($"The current time does not
            contain number {number}. Thread id:
            {Thread.CurrentThread.ManagedThreadId}");
        }
    });
}
```

Here is the call to `ExecuteParallelForEach` from the console application's `Main` method:

```
var numbers = new List<int> { 1, 3, 5, 7, 9, 0 };
var foreachExample = new ParallelForEachExample();
foreachExample.ExecuteParallelForEach(numbers);
```

Execute the program, and examine the console output. You should see that more than one thread was used to process the loop:

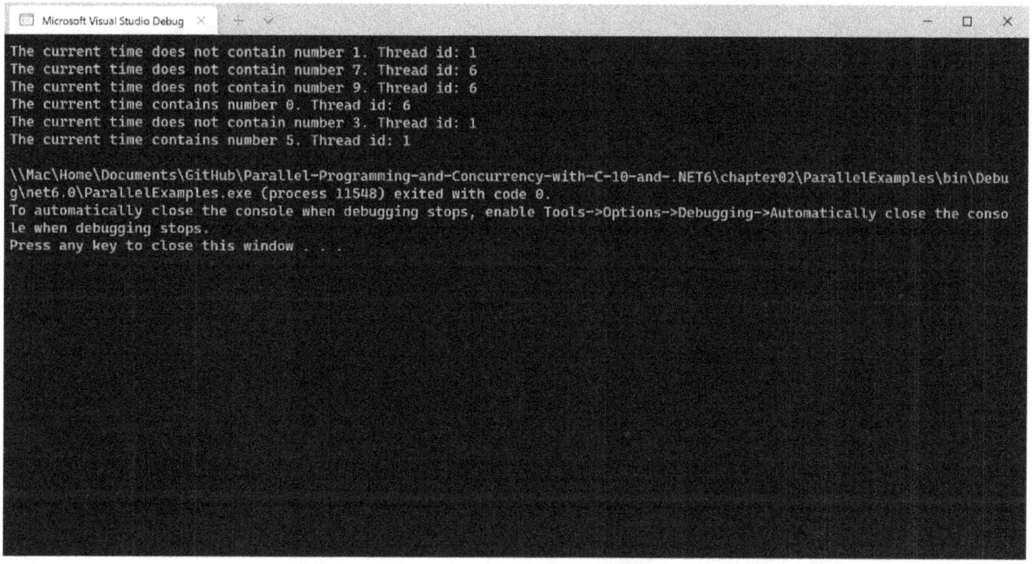

Figure 2.2 – Console output from a Parallel.ForEach loop

Next, we will wrap up this section on parallelism in .NET with an introduction to PLINQ.

Basics of Parallel LINQ

This section will look at one of the simplest ways to add some parallelism to your code. By adding the `AsParallel` method to your LINQ query, you can transform it into a PLINQ query, with the operations after `AsParallel` being executed on the thread pool when necessary. There are many factors to consider when deciding when to use PLINQ. We will discuss those in some depth in *Chapter 8*. For this example, we will introduce **PLINQ** inside a **LINQ** `Where` clause that checks whether each given integer is an even number. To help illustrate how PLINQ can impact sequences, `Task.Delay` is also introduced. Here is the complete `ParallelLinqExample` class implementation:

```
internal void ExecuteLinqQuery(IList<int> numbers)
{
    var evenNumbers = numbers.Where(n => n % 2 == 0);
```

```
        OutputNumbers(evenNumbers, "Regular");
    }
    internal void ExecuteParallelLinqQuery(IList<int> numbers)
    {
        var evenNumbers = numbers.AsParallel().Where(n =>
            IsEven(n));
        OutputNumbers(evenNumbers, "Parallel");
    }
    private bool IsEven(int number)
    {
        Task.Delay(100);
        return number % 2 == 0;
    }
    private void OutputNumbers(IEnumerable<int> numbers, string
        loopType)
    {
        var numberString = string.Join(",", numbers);
        Console.WriteLine($"{loopType} number string:
            {numberString}");
    }
```

In the Main method of your console application, add some code to pass a list of integers to both the ExecuteLinqQuery and ExecuteParallelLinqQuery methods:

```
var linqNumbers = new List<int> { 0, 1, 2, 3, 4, 5, 6, 7,
    8, 9, 10, 11, 12, 13, 14, 15, 16, 17, 18, 19, 20 };
var linqExample = new ParallelLinqExample();
linqExample.ExecuteLinqQuery(linqNumbers);
linqExample.ExecuteParallelLinqQuery(linqNumbers);
```

Examine the output, and you should see that the order of the numbers in the PLINQ sequence has changed:

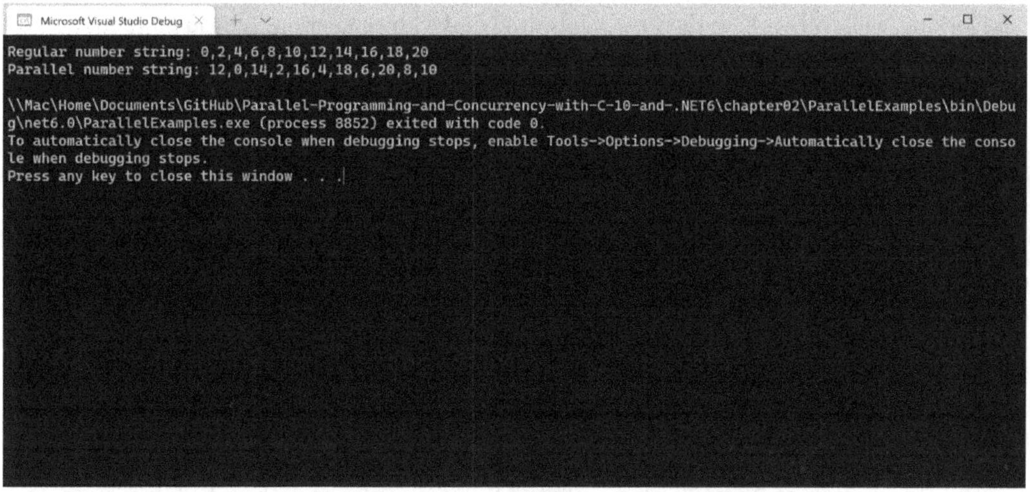

Figure 2.3 – Console output of the LINQ and PLINQ queries

We will explore more aspects of parallelism over several chapters in *Part 2, Parallel Programming and Concurrency with C#*, of this book. Let's shift gears and learn about some concurrency patterns in C#.

Introduction to concurrency

So, what is concurrency and how does it relate to parallelism in the context of C# and .NET? The terms are frequently used interchangeably, and if you think about it, they do have similar meanings. When multiple threads are executing in parallel, they are running concurrently. In this book, we will use the term concurrency when discussing patterns to follow when designing for managed threading. Additionally, we will discuss concurrency in the context of the concurrent collections that were introduced to .NET developers in .NET Framework 4.0. Let's start by learning about the concurrent collections in the System.Collections.Concurrent namespace.

.NET has several collections that have been created with thread safety built-in. These collections can all be found in the System.Collections.Concurrent namespace. In this section, we will introduce five of the collections. The remaining three are variations of Partitioner. These will be explored in *Chapter 9*, where we will work with each of the collection types through practical examples.

> **Note**
>
> While your code does not require the use of locks when using concurrent collections, different techniques for synchronization are being employed inside these collections. Some of them do use locking, while others have retry mechanisms to deal with contention. To read more about how concurrent collections handle contention, check out this Microsoft blog post on the .NET Parallel Programming blog: `https://devblogs.microsoft.com/pfxteam/faq-are-all-of-the-new-concurrent-collections-lock-free/`.

The first collection we will introduce is `ConcurrentBag<T>`.

ConcurrentBag<T>

The `ConcurrentBag<T>` collection is a concurrent collection intended to hold a collection of unordered objects. Duplicate values are allowed, as are null values when T is a nullable type. It makes an excellent thread-safe replacement for an array, `List<T>`, or other `IEnumerable<T>` instances where the ordering of items is not a requirement.

Internally, `ConcurrentBag<T>` stores a linked list of items for each thread adding items. As items are taken or peeked at from the collection, priority will be given to the internal list, which had items added by the current thread. Let's suppose thread 1 adds items A, B, and C and thread 2 adds items D, E, F, and G. If thread 1 calls `TryPeek` or `TryTake` four times, `ConcurrentBag<T>` will get items from the A, B, and C list first before taking items from the linked list containing items from thread 2.

The following list details the properties and methods of `ConcurrentBag<T>` that you are likely to use in most implementations:

- `Add(T)`: This adds an object to the bag.
- `TryPeek(out T)`: This tries to fetch a value from the bag with an `out` parameter but does not remove that item.
- `TryTake(out T)`: This attempts to fetch a value from the bag with an `out` parameter and removes it.
- `Clear()`: This clears all of the objects from the bag.
- `Count`: This returns the number of objects in the bag.
- `IsEmpty`: This returns a `bool` value indicating whether the bag is empty.
- `ToArray()`: This returns an array of objects of type T.

The `ConcurrentBag<T>` collection has two constructors. One constructor takes no parameters and simply creates a new empty bag. The other accepts an `IEnumerable<T>` type of object to be copied to the new bag.

Next, let's take a quick look at the `ConcurrentQueue<T>` collection.

ConcurrentQueue<T>

The .NET `ConcurrentQueue<T>` collection is similar in implementation to its thread-unsafe counterpart, `Queue<T>`. As such, it makes a great replacement for `Queue<T>` when managed threading is introduced to an existing code base. `ConcurrentQueue<T>` is a strongly typed list of objects that enforces **first in, first out (FIFO)** logic, which is the definition of a **queue**.

> **Note**
> **FIFO** logic is commonly seen in the manufacturing industry and in warehouse management software. When working with perishable goods, it is important to use your oldest raw materials first. Therefore, those pallets of goods that were the first to be put into the warehouse should be the first to be pulled when a pallet of that type is requested by the system.

These are the commonly used members of the `ConcurrentQueue<T>` type:

- `Enqueue(T)`: This adds a new object to the queue.
- `TryPeek(out T)`: This attempts to get the object at the front of the queue without removing it.
- `TryDequeue(out T)`: This tries to get the object at the front of the queue and removes it.
- `Clear()`: This clears the queue.
- `Count`: This returns the number of objects in the queue.
- `IsEmpty`: This returns a `bool` value indicating whether the queue is empty.
- `ToArray()`: This returns the objects in the queue as an array of type T.

Similar to the `ConcurrentBag<T>` collection, the `ConcurrentQueue<T>` collection has two constructors: one parameterless and one that takes an `Ienumerable<T>` type to populate the new queue. Next, let's introduce a similar collection: `ConcurrentStack<T>`.

ConcurrentStack<T>

`ConcurrentStack<T>` can be thought of as `ConcurrentQueue<T>`, but it uses **last in, first out (LIFO)**, or **stack**, logic instead of FIFO. The operations it supports are similar, but it has a `Push` method instead of `Enqueue`, and removing items uses a `TryPop` method instead of `TryDequeue`. Another advantage of the `ConcurrentStack<T>` collection is that it can add or remove multiple objects in one operation. These range operations are supported by using the `PushRange` and `TryPopRange` methods. The range operations take arrays of T as parameters.

`ConcurrentStack<T>` and `ConcurrentQueue<T>` in .NET 6 both implement the `IReadOnlyCollection<T>` interface. This means that once the collection has been created, it is read-only and cannot be reassigned or set to `null`. You can only add or remove items or use `Clear()` to empty the collection.

Let's move on to one of the most powerful concurrent collections, `BlockingCollection<T>`.

BlockingCollection<T>

`BlockingCollection<T>` is a thread-safe collection of objects that implements several interfaces, including `IProducerConsumerCollection<T>`. The `IProducerConsumerCollection<T>` interface provides a set of members intended to support applications that need to implement the **producer/consumer pattern**.

Note

The producer/consumer pattern is a concurrency design pattern where a set of data is concurrently fed by one or more producer threads. At the same time, there are one or more consumer threads monitoring and fetching the data being produced to consume and process it concurrently. The `BlockingCollection<T>` collection is the data store in this producer/consumer pattern. You can read more about producer/consumer on Wikipedia at `https://en.wikipedia.org/wiki/Producer%E2%80%93consumer_problem`.

`BlockingCollection<T>` has several methods and properties that assist in a producer/consumer workflow. You can indicate that the producer process is done adding items to the collection by calling the `CompleteAdding` method. Once this method is called, no more items can be added to the collection with the `Add` or `TryAdd` methods. If you plan to use `CompleteAdding` in your workflow, it is best to always use `TryAdd` and check the Boolean result when adding objects to the collection. If the collection has been marked as complete for adding, calling `Add` will throw `InvalidOperationException`. Additionally, you can check the `IsAddingCompleted` property to find out whether `CompleteAdding` has already been called.

Items are removed from `BlockingCollection<T>` by a consumer process with the `Take` or `TryTake` methods. Again, it is safer to use `TryTake` to avoid any exceptions when the collection is empty. If `CompleteAdding` has been called and all objects have been removed from the collection, the `IsCompleted` property will return true.

We will walk through a real-world producer/consumer implementation in *Chapter 9 .NET*. Now, let's move on to our final concurrent collection in this section, `ConcurrentDictionary<T>`.

ConcurrentDictionary<TKey, TValue>

As you have probably guessed, `ConcurrentDictionary<TKey, TValue>` is a great replacement for `Dictionary<TKey, TValue>` when working with managed threading. Both collections implement the `IDictionary<TKey, TValue>` interface. The concurrent version of this collection adds the following methods for working with data concurrently:

- `TryAdd`: This tries to add a new key/value pair to the dictionary and returns a Boolean value indicating whether the object was successfully added to the dictionary. If the key already exists, the method will return `false`.

- `TryUpdate`: This operation passes a key along with the existing and new values for the item. It will update the existing item to the new value if it exists in the dictionary with the existing value provided. The Boolean value that is returned indicates whether the object was successfully updated in the dictionary.

- `AddOrUpdate`: This method will add or update an item in the dictionary based on whether the key exists and uses an update delegate to perform any logic based on the current and new values for the item.

- `GetOrAdd`: This method will add an item if the key does not already exist. If it does exist, the existing value is returned.

These are the most important and common concurrent collections in .NET to understand. We will cover some examples of each and learn about more of the collections in `System.Collections.Concurrent` later, but this section should provide a solid base for understanding what is to come.

In the next section, we will introduce the C# `async` and `await` keywords that were added to C# 5.0.

Basics of async and await

When the TPL was introduced in .NET Framework 4.5, C# 5.0 also added language support for task-based asynchronous programming with the `async` and `await` keywords. This immediately became the default method of implementing asynchronous workflows in C# and .NET. Now, 10 years later, `async`/`await` and the TPL have become an integral part of building robust, scalable .NET applications. You might be wondering why it is so important to adopt async programming in your applications.

Understanding the async keyword

There are many reasons for writing async code. If you're writing server-side code on a web server, using async allows the server to handle additional requests while your code is awaiting a long-running operation. On a client application, freeing the UI thread to perform other operations with async code allows your UI to remain responsive to users.

Another important reason to adopt async programming in .NET is that many third-party and open source libraries are using `async`/`await`. Even .NET itself is exposing more APIs as async in every release, especially those involving IO operations: networking, file, and database access.

Let's try writing your first async method with C#.

Writing an async method

Creating and consuming an async method is easy. Let's try a simple example with a new console application:

1. Create a new console application in Visual Studio and name it **AsyncConsoleExample**.

2. Add a class to the project, named `NetworkHelper`, and add the following methods to the class:

```csharp
internal async Task CheckNetworkStatusAsync()
{
    Task t = NetworkCheckInternalAsync();
    for (int i = 0; i < 8; i++)
    {
        Console.WriteLine("Top level method
            working...");
        await Task.Delay(500);
    }
    await t;
}
private async Task NetworkCheckInternalAsync()
{
    for (int i = 0; i < 10; i++)
    {
        bool isNetworkUp = System.Net.
            NetworkInformation.NetworkInterface
                .GetIsNetworkAvailable();
        Console.WriteLine($"Is network available?
            Answer: {isNetworkUp}");
        await Task.Delay(100);
    }
}
```

There are a few things to point out in the preceding code. Both methods have an `async` modifier, indicating that they will be awaiting some work and will run asynchronously. Inside the methods, we are using the `await` keyword with the calls to `Task.Delay`. This will ensure that no code after this point will execute until the awaited method has been completed. However, during this time, the active thread can be released to perform other work elsewhere.

Finally, look at the call to `NetworkCheckInternalAsync`. Instead of awaiting this call, we are capturing the returned `Task` instance in a variable named `t`, and we don't `await` it until after the `for` loop. This means that the `for` loops in both methods will run concurrently. If we had, instead, awaited `NetworkCheckInternalAsync`, its `for` loop would have been completed before the `for` loop in `CheckNetworkStatusAsync` could begin.

3. Next, replace the code in `Program.cs` with the following:

```
using AsyncConsoleExample;
Console.WriteLine("Hello, async!");
var networkHelper = new NetworkHelper();
await networkHelper.CheckNetworkStatusAsync();
Console.WriteLine("Main method complete.");
Console.ReadKey();
```

We are awaiting the call to `CheckNetworkStatusAsync`. This is possible because the default `Main` method in a .NET 6 console application is `async` by default. If you try to `await` something in a method that is not marked as `async`, you will get a compiler error. We will explore some of the options you can use when you must call async methods from existing non-async code in *Chapter 5*.

4. Finally, run the application and examine the output:

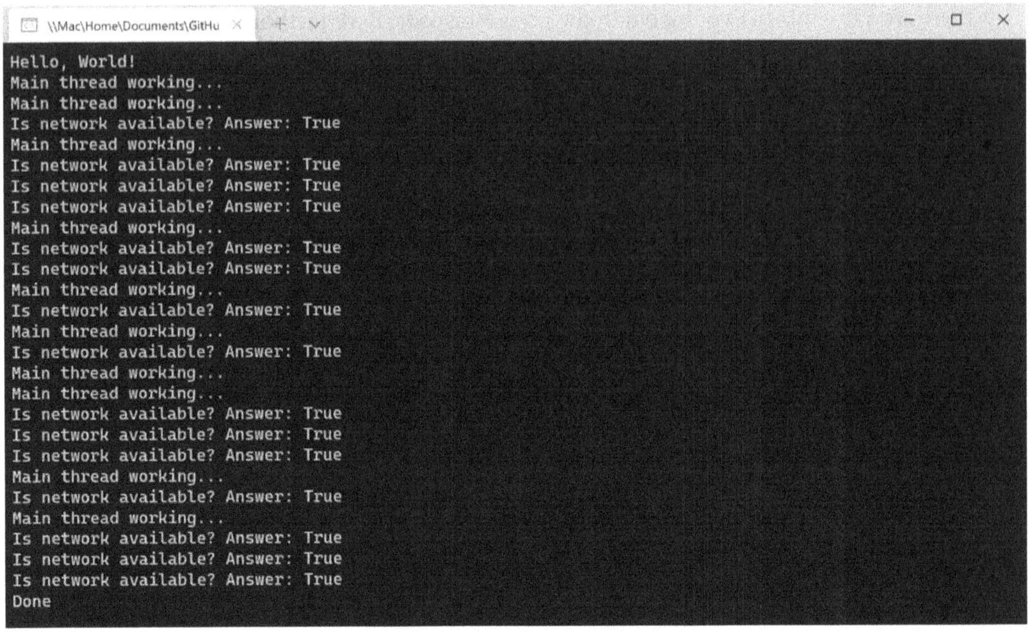

Figure 2.4 – Console output for the async sample application

As expected, capturing the async method's result allowed the two loops to run concurrently. Try awaiting the call to `NetworkCheckInternalAsync` and see how the output changes. You should see that all the output from the private method will appear before the output from the `for` loop in `CheckNetworkStatusAsync` begins.

This was a brief introduction to the world of async programming with C#. We'll be working with it quite a lot throughout the rest of this book. Let's wrap things up by discussing how to choose which of these options to leverage when building a new project or enhancing an existing application.

Choosing the right path forward

Now that you have been introduced to some advanced managed threading concepts, parallel programming, concurrent collections, and the async/await paradigm, let's discuss how they all fit together in the real world. Choosing the right path forward with multithreaded development in .NET will usually involve more than one of these concepts.

When working with .NET 6, you should usually choose to create `async` methods in your projects. The reasons discussed in this chapter are compelling. Asynchronous programming keeps both client and server applications responsive, and `async` is used extensively throughout .NET itself.

Some of the `Parallel` class operations can be leveraged when your code needs to process a set of items quickly and the underlying code doing the processing is thread-safe. This is one place where concurrent collections can be introduced. If any parallel or async operations are manipulating shared data, the data should be stored in one of the .NET concurrent collections.

If you are working with existing code, often, the most prudent path is to limit how much multithreaded code is added. Legacy projects such as these are a great place to incrementally add some `ThreadPool` or `Parallel` operations and test the results. It is important to test the application functionally and for performance. Performance testing tools for managed threading will be covered in *Chapter 10*.

This preliminary guidance will help you get an idea of where you can boost your applications' performance with managed threading. We will build on your learning and this guidance throughout the rest of the book. Let's wrap up and discuss what you have learned in this chapter.

Summary

In this chapter, we started by looking at a brief history of C#, .NET, and managed threading. We discussed how Microsoft has added features for asynchronous and parallel programming over the last 20 years. Next, we took a tour of parallel programming with .NET, concurrent collections, and asynchronous development with C#. Finally, we examined when you might choose each of these concepts for your own applications and why you will often choose more than one of them. You will be able to take what you learned in this chapter and start thinking about practical applications of managed threading in your day-to-day work.

In the next chapter, we will take what you have learned so far and discuss some of the best practices for the practical application of the concepts.

Questions

1. Which class in .NET manages the thread pool threads available to your application?
2. In which version of C# were the `async` and `await` keywords introduced?
3. In which version of .NET was the TPL introduced?
4. In which version of .NET Core was `IAsyncEnumerable` introduced?
5. What type should every `async` method return?
6. Which concurrent collection would you choose to replace `Dictionary<TKey, TValue>` in a multithreaded scenario?
7. Which concurrent collection is frequently used with the producer/consumer design pattern in .NET?
8. Which parallel feature in .NET features the `AsParallel` method?

Best Practices for Managed Threading

When building applications that leverage parallelism and concurrency, developers need to be aware of some best practices regarding integrating managed threading concepts. This chapter will assist in this capacity. We will cover important concepts such as working with static data, avoiding deadlocks, and exhausting managed resources. These are all areas that can lead to unstable applications and unexpected behavior.

In this chapter, you will learn the following concepts:

- Handling static objects
- Managing deadlocks and race conditions
- Threading limits and other recommendations

By the end of this chapter, you will have the knowledge to avoid the most common managed threading pitfalls.

Technical requirements

To follow along with the examples in this chapter, the following software is recommended for Windows developers:

- Visual Studio 2022 version 17.0 or later
- .NET 6

While these are recommended, if you have .NET 6 installed, you can use your preferred editor. For example, Visual Studio 2022 for Mac on macOS 10.13 or later, JetBrains Rider, or Visual Studio Code will work just as well.

All the code examples for this chapter can be found on GitHub at `https://github.com/ PacktPublishing/Parallel-Programming-and-Concurrency-with-C- sharp-10-and-.NET-6/tree/main/chapter03`.

We will get started by discussing some best practices for handling static data in .NET.

Handling static objects

When working with static data in .NET, there are some important things to understand when it comes to managed threading.

Static data and constructors

One important item to understand about accessing static data from managed threads relates to constructors. Before a static member of any class can be accessed, its **static constructor** must first finish running. The runtime will block thread execution until the static constructor has run to ensure that all required initialization has finished.

If you are using static objects within your own code base, you will know which classes have static constructors and can control the complexity of the logic inside them. When the static data is outside of your control, inside a third-party library or .NET itself, things may not be so clear.

Let's try a quick example to illustrate the potential delays that can be encountered in this scenario.

1. Start by creating a new .NET console application in Visual Studio named `ThreadingStaticDataExample`.

2. Add a new class to the project named `WorkstationState` with the following static members:

    ```
    internal static string Name { get; set; }
    internal static string IpAddress { get; set;}
    internal static bool IsNetworkAvailable { get; set; }
    internal static DateTime? NetworkConnectivity
        LastUpdated { get; set; }

    static WorkstationState()
    {
        Name = Dns.GetHostName();
        IpAddress = GetLocalIPAddress(Name);
    ```

```
    IsNetworkAvailable = NetworkInterface
        .GetIsNetworkAvailable();
    NetworkConnectivityLastUpdated = DateTime.UtcNow;
    Thread.Sleep(2000);
}


private static string GetLocalIPAddress
    (string hostName)
{
    var hostEntry = Dns.GetHostEntry(hostName);
    foreach (var address in hostEntry.AddressList
                        .Where(a => a.AddressFamily ==
                        AddressFamily.InterNetwork))
    {
        return address.ToString();
    }
    return string.Empty;
}
```

This class will hold some information about the current workstation, including the host name, local IP address, and whether the network is currently available. The private GetLocalIpAddress method fetches the local IP based on a provided host name.

There is a static constructor for WorkstationState that sets the initial property data and injects a delay of two seconds with a Thread.Sleep call. This will help us simulate the application fetching some other network information that takes some time to retrieve on a slow network connection.

3. Next, add a class named WorkstationHelper. This class will contain an async method to update the static IsNetworkAvailable and NetworkConnectivityLastUpdated properties in WorkstationState and return the value of IsNetworkAvailable to the caller:

```
internal async Task<bool> GetNetworkAvailability()
{
    await Task.Delay(100);
    WorkstationState.IsNetworkAvailable =
        NetworkInterface.GetIsNetworkAvailable();
    WorkstationState.NetworkConnectivityLastUpdated =
        DateTime.UtcNow;
```

```
            return WorkstationState.IsNetworkAvailable;
    }
```

There is also a `Task.Delay` call being awaited if you would like to call this in a loop and experiment by varying the injected delay.

4. Finally, update `Program.cs` to call `GetNetworkAvailability` and update the console output with the connectivity, host name, and IP address:

```
using ThreadingStaticDataExample;
Console.WriteLine("Hello, World!");
Console.WriteLine($"Current datetime:
    {DateTime.UtcNow}");
var helper = new WorkstationHelper();
await helper.GetNetworkAvailability();

Console.WriteLine($"Network availability last updated
  {WorkstationState.NetworkConnectivityLastUpdated}
    for computer {WorkstationState.Name} at IP
      {WorkstationState.IpAddress}");
```

5. Run the program and examine the output. You can see that there is a two second delay between the times in the two `Console.WriteLine` calls injected by the static constructor:

```
Hello, World!
Current datetime: 2/12/2022 4:07:13 PM
Network availability last updated 2/12/2022 4:07:15 PM for
computer ALVINASHCRABC3A at IP 10.211.55.3
```

Static constructors are one aspect of static data to keep in mind when working with managed threading. A more common issue is controlling concurrent read/write access to static objects across threads.

Controlling shared access to static objects

When it comes to static data, the best practice is to avoid using it whenever possible. In general, it makes your code less testable, less scalable, and more prone to unexpected behavior when working with concurrency. However, there are times when static data cannot be avoided. You may be working with a legacy code base, where refactoring the code to remove statics can be risky or too large an effort to undertake. Static classes can also be useful when data rarely changes, or when the classes are stateless.

For cases where static objects are unavoidable, some precautions can be taken. Let's review some of them and discuss the merits of each, starting with locking mechanisms.

Locks

In *Chapter 1,* we discussed some strategies for locking objects for shared use. **Locks** are even more important when working with static variables because of the chance of concurrent access increases with the increased scope of the object.

The simplest way of preventing concurrent access to an object from multiple threads is to enclose any code that accesses it with a lock. Let's modify the code in WorkstationHelper to prevent multiple calls to GetNetworkActivity from writing to WorkstationState properties concurrently:

```
internal class WorkstationHelper
{
    private static object _workstationLock = new object();
    internal async Task<bool> GetNetworkAvailability()
    {
        await Task.Delay(100);
        lock( _workstationLock)
        {
            WorkstationState.IsNetworkAvailable =
                NetworkInterface.GetIsNetworkAvailable();
            WorkstationState.NetworkConnectivityLastUpdated
                = DateTime.UtcNow;
        }
        return WorkstationState.IsNetworkAvailable;
    }
}
```

We have added a private static _workstationLock object, and we are using it as part of the lock block enclosing the writes to WorkstationState properties. If GetNetworkAvailability were now used in a Parallel.ForEach or some other concurrent operation, only one thread could enter that lock block at a time.

You can use any of the locking mechanisms that were discussed in *Chapter 1.* Choose the feature that works best for your scenario. Another .NET feature you can leverage with static members is the ThreadStatic attribute.

ThreadStatic attribute

The ThreadStatic attribute can be added to a static field to indicate that a separate static instance of the object should be created for each thread. The ThreadStatic attribute should only be used when this is the desired behavior, and it is well documented in your code. It can produce unexpected results when used improperly.

Fields marked as ThreadStatic should not have their data initialized in a constructor, as the initialization will only apply to the current thread. The value on all other threads will be null or the default value for that type.

If you applied the ThreadStatic attribute to the NetworkConnectivityLastUpdated property of WorkstationState and call WorkstationHelper. GetNetworkAvailability thirty times in a Parallel.For loop, the value read in Program.cs at the end may or may not be the last value written to one of the static instances. The variable in Program.cs will contain the last value written from the main thread inside the Parallel.For loop.

1. To try it for yourself, add the ThreadStatic attribute to NetworkConnectivityLastUpdated and make it an internal field instead of a property. The attribute cannot be applied to properties:

    ```
    [ThreadStatic]
    internal static DateTime?
        NetworkConnectivityLastUpdated;
    ```

2. Then update Program.cs to use a Parallel.For loop:

    ```
    using ThreadingStaticDataExample;
    Console.WriteLine("Hello, World!");
    Console.WriteLine($"Current datetime:
        {DateTime.UtcNow}");
    var helper = new WorkstationHelper();

    Parallel.For(1, 30, async (x) =>
    {
        await helper.GetNetworkAvailability();
    });

    Console.WriteLine($"Network availability last updated
        {WorkstationState.NetworkConnectivityLastUpdated}
            for computer {WorkstationState.Name} at IP
                {WorkstationState.IpAddress}");
    ```

The time between the date/time values in the output will now vary each time you run the program because the final value written to the console may not be the final value across all threads.

While `ThreadStatic` should be applied only in scenarios where instances per thread are necessary, another pattern similar in application to statics is the **singleton**. Let's discuss the use of singletons in a multithreaded application.

Working with singletons

The singleton pattern is an object **design pattern** that only allows a single instance of itself to be created. This design pattern is one of the most common and is known by most .NET developers. Every mainstream **dependency injection (DI)** framework allows registered types to be registered as singletons. The container will only create one instance for each of these types, providing the same instances every time the type is requested.

We can manually create a singleton for our `WorkstationState` with a `lock` and a little extra code. This is the `WorkstationStateSingleton`:

```
public class WorkstationStateSingleton
{
    private static WorkstationStateSingleton?
        _singleton = null;
    private static readonly object _lock = new();
    WorkstationStateSingleton()
    {
        Name = Dns.GetHostName();
        IpAddress = GetLocalIPAddress(Name);
        IsNetworkAvailable =
            NetworkInterface.GetIsNetworkAvailable();
        NetworkConnectivityLastUpdated =
            DateTime.UtcNow;
    }

    public static WorkstationStateSingleton Instance
    {
        get
        {
            lock (_lock)
            {
                if (_singleton == null)
                {
                    _singleton = new
```

```
                          WorkstationStateSingleton();
                }
                return _singleton;
            }
        }
    }
    ...
}
```

The complete implementation of the class can be found in the GitHub repository referenced in the *Technical requirements* section of this chapter. Look at the `ThreadingStaticDataExample` in the `chapter3` folder.

There are two steps taken to make this a singleton. First, the constructor is private so only the `WorkstationStateSingleton` can create an instance of itself. Second, a static `Instance` method is created. It returns the `_singleton` instance of itself if it is not `null`. Otherwise, it creates the instance to return. Surrounding this code with the `_lock` ensures that the instances are not created twice on different concurrent threads.

A singleton presents the same challenges as a static class. All shared data should be protected by locks if they can be accessed concurrently by managed threads. The added challenge with singletons that are registered in a DI container is that a `lock` object, `Mutex`, or another mechanism must be declared at the same scope as the container. This will ensure that all data that can potentially use the singleton can also enforce the same lock.

> **Note**
>
> Please note that the use of singletons is generally not considered a good practice today. For this reason, many developers consider them an anti-pattern. However, it is important to understand them and how existing singletons in your code may be impacted by multithreaded code.

Deadlocks are one of the pitfalls of aggressive locking. Aggressive locking is when you are locking uses of an object in many parts of the code that could be executing in parallel. In the next section, we will discuss deadlocks and **race conditions** in managed threading.

Managing deadlocks and race conditions

As with many tools at a developer's disposal, misusing features of managed threading can have adverse impacts on your applications at runtime. Deadlocks and race conditions are two scenarios that can be created because of multithreaded programming:

- A **deadlock** happens when multiple threads are trying to lock the same resource and as a result, cannot continue executing.

- **Race conditions** happen when multiple threads are proceeding toward updating a particular routine, and a correct outcome is dependent on the order in which they execute it.

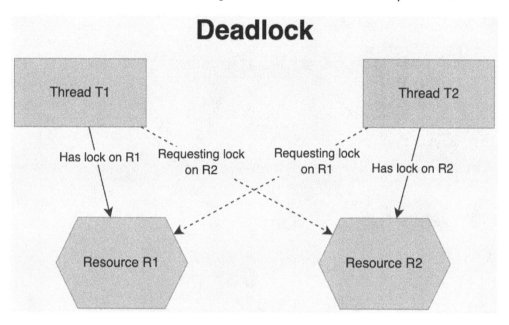

Figure 3.2 – Two threads in contention for the same resources, causing a deadlock

First, let's discuss deadlocks and some techniques for avoiding them.

Mitigating deadlocks

It is critical to avoid deadlocks in your applications. If one of the threads involved in a deadlock is the application's UI thread, it will cause the application to freeze. When only non-UI threads are deadlocked, it can be harder to diagnose the problem. Deadlocked thread pool threads will prevent an application from closing, but deadlocked background threads will not.

Well-instrumented code is essential in debugging problems when they occur in a production environment. If the issue can be reproduced in your own development environment, stepping through the code with the Visual Studio debugger is the fastest way to find the source of a deadlock. We will discuss debugging techniques in detail in *Chapter 10*.

One of the easiest ways to create a deadlock is through recursion or nested methods that try to acquire a lock on the same resource. Look at the following code:

```csharp
private object _lock = new object();
private List<string> _data;
public DeadlockSample()
{
    _data = new List<string> { "First", "Second",
        "Third" };
}

public async Task ProcessData()
{
    lock (_lock)
    {
        foreach(var item in _data)
        {
            Console.WriteLine(item);
        }
        await AddData();
    }
}

private async Task AddData()
{
    lock (_lock)
    {
        _data.AddRange(GetMoreData());
        await Task.Delay(100);
    }
}
```

The ProcessData method is locking the _lock object and processing with _data. However, it is calling AddData, which also tries to acquire the same lock. This lock will never become available, and the process will be deadlocked. In this case, the problem is apparent. What if AddData is called from multiple places or some Parallel.ForEach any loops are involved in the parent code? Some of the parent code uses _data and acquire a lock, but some do not. This is a case where non-blocking read locks in the ReaderWriterLockSlim can help prevent deadlocks.

Another way to prevent deadlocks is by adding a timeout to the lock attempt with `Monitor.TryEnter`. In this example, the code will time out if a lock cannot be acquired within one second:

```
private void AddDataWithMonitor()
{
    if (Monitor.TryEnter(_lock, 1000))
    {
        try
        {
            _data.AddRange(GetMoreData());
        }
        finally
        {
            Monitor.Exit(_lock);
        }
    }
    else
    {
        Console.WriteLine($"AddData: Unable to acquire
            lock. Stack trace: {Environment.StackTrace}");
    }
}
```

Logging any failures to acquire locks can help to pinpoint possible sources of deadlocks in your code so you can rework the code to avoid them.

Next, let's examine how race conditions can occur in multithreaded applications.

Avoiding race conditions

A race condition occurs when multiple threads are reading and writing the same variables simultaneously. Without any locks in place, the outcome can be wildly unpredictable. Some operations can be overwritten by other parallel threads' results. Even with locks in place, the order of two thread operations can change the result. Here is a simple example without locks that performs some addition and multiplication in parallel:

```
private int _runningTotal;
public void PerformCalculationsRace()
{
    _runningTotal = 3;
```

```
    Parallel.Invoke(() => {
        AddValue().Wait();
    }, () => {
        MultiplyValue().Wait();
    });
    Console.WriteLine($"Running total is {_runningTotal}");
}
private async Task AddValue()
{
    await Task.Delay(100);
    _runningTotal += 15;
}
private async Task MultiplyValue()
{
    await Task.Delay(100);
    _runningTotal = _runningTotal * 10;
}
```

We all know that when combining addition and multiplication, the order of operations is important. If the two operations are processed sequentially, the two results could be either 180 or 45, but if both AddValue and MultiplyValue read the initial value of 3 before performing their respective operations, the last method to complete will write either 18 or 30 as the final value of _runningTotal.

If you want to ensure that multiplication happens before addition, the PerformCalculations method can be rewritten to use the ContinueWith method on the Task returned from MultiplyValue:

```
public async Task PerformCalculations()
{
    _runningTotal = 3;
    await MultiplyValue().ContinueWith(async (Task) => {
        await AddValue();
        });
    Console.WriteLine($"Running total is {_runningTotal}");
}
```

This code will always multiply before adding and will always finish with _runningTotal equaling 45. Using async and await throughout the code ensures that the UI or service process remains responsive while using threads from the thread pool as needed.

The Interlocked class discussed in the previous chapter can also be used to perform mathematic operations on shared resources. Interlocked.Add and Interlocked.Exchange can perform thread-safe operations on the _runningTotal variable in parallel. Here is the original Parallel.Invoke example modified to use Interlocked methods with _runningTotal:

```csharp
public class InterlockedSample
{
    private long _runningTotal;
    public void PerformCalculations()
    {
        _runningTotal = 3;
        Parallel.Invoke(() => {
            AddValue().Wait();
        }, () => {
            MultiplyValue().Wait();
        });
        Console.WriteLine($"Running total is
            {_runningTotal}");
    }
    private async Task AddValue()
    {
        await Task.Delay(100);
        Interlocked.Add(ref _runningTotal, 15);
    }


    private async Task MultiplyValue()
    {
        await Task.Delay(100);
        var currentTotal = Interlocked.Read(ref
            _runningTotal);
        Interlocked.Exchange(ref _runningTotal,
            currentTotal * 10);
    }
}
```

The two operations could still perform in different orders, but the uses of `_runningTotal` are now locked and thread-safe. The **Interlocked** class is more efficient than using a lock statement and will yield greater performance for simple changes like these.

It is important to guard all shared resources when performing concurrent operations in your code. By creating a well-designed locking strategy, you will achieve the best possible performance while maintaining thread safety in your application. Let's finish up this chapter with some guidance around threading limits.

Threading limits and other recommendations

So, it sounds like using multiple threads can really speed up your application's performance. You should probably start replacing all your `foreach` loops with `Parallel.ForEach` loop and start calling all your services and helper methods on thread pool threads, right? Are there any limits and what are they? Well, when it comes to threading, there absolutely are limits.

The number of threads that can execute simultaneously is limited by the number of processors and processor cores on the system. There is no way around hardware limitations, as the CPU (or virtual CPU when running on a virtual machine) can only run so many threads. In addition, your application must share these CPUs with other processes running on the system. If your CPU has four cores, it is actively running five other applications, and your program is trying to execute a process with multiple threads, the system is not likely to accept more than one of your threads at a time.

The .NET thread pool is optimized to handle different scenarios based on the number of threads available, but you can do some things to guard against taxing the system. Some parallel operations such as `Parallel.ForEach` can limit how many threads the loop will try to use. You can provide a `ParallelOptions` object to the operation and set the `MaxDegreeOfParallelism` option. By default, the loop will use as many threads as the scheduler will provide.

You can ensure that the maximum does not exceed half the number of available cores on the system with the following implementation:

```
public void ProcessParallelForEachWithLimits
    (List<string> items)
{
    int max = Environment.ProcessorCount > 1 ?
                Environment.ProcessorCount / 2 : 1;
    var options = new ParallelOptions
    {
        MaxDegreeOfParallelism = max
    };
    Parallel.ForEach(items, options, y => {
```

```
        // Process items
    });
}
```

PLINQ operations can also limit the max degree of parallelism with the WithDegreeOfParallelism extension method:

```
public bool ProcessPlinqWithLimits(List<string> items)
{
    int max = Environment.ProcessorCount > 1 ?
        Environment.ProcessorCount / 2 : 1;
    return items.AsParallel()
        .WithDegreeOfParallelism(max)
        .Any(i => CheckString(i));
}


private bool CheckString(string item)
{
    return !string.IsNullOrWhiteSpace(item);
}
```

An application can also adjust the thread pool maximum values, if necessary. By calling ThreadPool.SetMaxThreads, you can change the maximum values for workerThreads and completionPortThreads. completionPortThreads is the number of async I/O threads on the thread pool. It is usually not required to change these values, and there are some limits to the values you can set. The maximum cannot be set to less than the number of cores on the system or less than the current minimum values on the thread pool. You can query the current minimums with ThreadPool.GetMinThreads. Here is an example of how to safely set the maximum thread values to values greater than the current minimums:

```
private void UpdateThreadPoolMax()
{
    ThreadPool.GetMinThreads(out int workerMin, out int
        completionMin);
    int workerMax = GetProcessingMax(workerMin);
    int completionMax = GetProcessingMax(completionMin);
    ThreadPool.SetMaxThreads(workerMax, completionMax);
}
private int GetProcessingMax(int min)
```

```
{
    return min < Environment.ProcessorCount ?
                Environment.ProcessorCount * 2 :
                min * 2;
}
```

There are some other general guidelines to follow regarding the number of threads to assign to an operation in your application. Try to avoid assigning multiple threads to operations that share a resource. For example, if you have a service that logs activity to a file, you should not assign more than one background worker to do the logging. The blocking file I/O operations will prevent the second thread from writing until the first one is complete. You are not gaining any efficiency in this case.

If you find yourself adding extensive locking to objects in your application, you are either using too many threads or the task distribution needs to be changed to reduce contention for resources. Try to divide threaded task responsibility by the types of data being consumed. You might have many parallel tasks calling services to fetch data, but only one or two threads are needed to process the data once it is returned.

You may have heard the term **thread starvation**. This usually happens when too many threads are blocking or waiting for resources to become available. There are some common scenarios where this happens:

- **Locks**: There are too many threads competing for the same locked resources. Analyze your code to determine how to reduce contention.

- **No async/await**: When working with **ASP.NET Core**, all controller methods should be marked as `async`. This allows the webserver to serve other requests while yours are waiting for operations to complete.

- **Too much threading**: Creating too many thread pool threads will result in more idle threads waiting to be processed. It also increases the likelihood of thread contention and starvation.

Avoid these practices, and .NET will do its best to manage the thread pool to serve your application and others on the system.

Finally, do not use `Thread.Suspend` and `Thread.Resume` trying to control the sequence of operations across multiple threads. Instead, leverage other techniques discussed in this chapter, including locking mechanisms and `Task.ContinueWith`.

We have covered plenty of best practices for managed threading in this chapter. Let's wrap up by reviewing what we have learned.

Summary

In this chapter, we discussed some best practices to follow when working with managed threads in C# and .NET. We started by creating some examples of how to manage and process static data in a multithreaded application. The examples illustrated how to leverage locks, work with singletons, and how static constructors can impact performance when working with static data. Next, we explored some techniques for avoiding deadlocks and race conditions. Both pitfalls can be avoided if you design your algorithms to minimize the need for locking. Finally, we looked at some features of .NET that can adjust the limits of several parallel and thread pool operations.

At this point, you are well prepared to start using managed threads responsibly in your .NET projects. For some further reading on best practices with managed threading, you can check out some recommendations on Microsoft Docs: `https://docs.microsoft.com/en-us/dotnet/standard/threading/managed-threading-best-practices`.

In the next chapter, *Chapter 4*, you will learn how to leverage parallelism and concurrency to keep your application responsive and pick up some best practices for updating the UI from a non-UI thread.

Questions

1. Which design pattern models how to create an object that only has one instance?
2. What .NET attribute will cause a static field to have one instance per thread?
3. What is a threading deadlock?
4. Which method on the `Monitor` class can be used to specify a timeout when trying to access a locked resource?
5. Which lightweight class can be used to lock value types for atomic operations?
6. Which thread-safe operation can be used to add two integers?
7. What option can be set on a `Parallel.For` or `Parallel.ForEach` loop to limit the number of threads used?
8. How can you limit the number of threads used in a PLINQ query?
9. What is the name of the method to find the current minimum thread values on the thread pool?

User Interface Responsiveness and Threading

One of the main reasons to introduce threading concepts to a project is the desire to keep an application responsive to user input. Accessing data through services, a database, or the filesystem can introduce delays, and the **user interface** (**UI**) should remain responsive. The real-world examples in this chapter will provide valuable options for ensuring UI responsiveness in your .NET client applications.

In this chapter, we will do the following:

- Leveraging background threads
- Using the thread pool
- Updating the UI thread without exceptions

By the end of this chapter, you will understand how to take advantage of parallelism and concurrency to keep your client applications responsive and performant.

Technical requirements

To follow along with the examples in this chapter, the following software is recommended for Windows users:

- Visual Studio 2022 version 17.0 or later
- .NET 6

While these are recommended, if you have .NET 6 installed, you can use your preferred editor. For example, Visual Studio 2022 for Mac on macOS 10.13 or later, JetBrains Rider, or Visual Studio Code will work just as well.

All the code examples for this chapter can be found on GitHub at `https://github.com/ PacktPublishing/Parallel-Programming-and-Concurrency-with-C-sharp- 10-and-.NET-6/tree/main/chapter04`.

Let's get started by discussing how background threads can be used to perform non-critical tasks without impacting UI performance.

Leveraging background threads

In *Chapter 1*, we learned how to create background threads and discussed some of their uses. Background threads have a lower priority than the primary thread of the process and other thread pool threads. In addition, active background threads will not prevent the user or the system from terminating the application.

This means that background threads are perfect for tasks such as the following:

- Writing log and analytics data
- Monitoring network or filesystem resources
- Reading data into the application

Do not use background threads for critical application operations such as the following:

- Saving application state
- Performing database transactions
- Application data processing

A good rule to follow when deciding whether some work can be processed by a background thread is to ask yourself whether abruptly interrupting the work to close the application would risk the data integrity of the system. So, how do you know whether you are creating a background or foreground thread?

Which threads are background threads?

We have learned that a thread can be explicitly created as a background thread by setting its `IsBackground` property to `true`. All other threads created by calling a `Thread` constructor are foreground threads by default. The application's primary (or main) thread is a foreground thread. All `ThreadPool` threads are background threads. This includes all asynchronous operations started by the **Task Parallel Library** (**TPL**).

So, if all task-based operations such as `async` methods are executing on background threads, should you avoid using them for saving important application data? Will .NET allow your application to close while these `async` / `await` operations are in process? If there is a foreground thread awaiting an `async` operation, the application will not terminate until the operation is complete. If you do not use `await`, or you start an operation on the thread pool with `Task.Run`, it is possible for the application to terminate normally before the actions have finished.

The great thing about using `await` with your `async` methods is the flexibility you gain in controlling the flow of execution while keeping the UI responsive. Let's discuss `async` and `await` in client applications and create an example of a **Windows Presentation Foundation (WPF)** application that loads data from multiple sources.

Using async, await, tasks, and WhenAll

Using `async` and `await` in your code is the easiest way to introduce some background work using `ThreadPool`. An asynchronous method must be decorated with the `async` keyword and will return a `System.Threading.Tasks.Task` type instead of a `void` return.

> **Note**
>
> Async methods return `Task` so the calling method can await the result of the method. If you were to create an `async` method with a `void` return type, it could not be awaited, and the calling code would continue processing subsequent code before the `async` method had completed. It is important to note that only event handlers should be declared as `async` with a `void` return type.

If the method returns `string`, then the `async` equivalent will return a `Task<string>` generic type. Let's look at examples of each:

```
private async Task ProcessDataAsync()
{
    // Process data here
}
private async Task<string> GetStringDataAsync()
{
    string stringData;
    // Build string here
    ...
    return stringData;
}
```

When you call an `async` method, there are two common patterns to follow.

- First, you can await the call and set the return type to a variable of the type returned inside the method:

```
await ProcessDataAsync();
string data = await GetStringDataAsync();
```

- The second option is to use `Task` variables when invoking the methods and await them later:

```
Task dataTask = ProcessDataAsync();
Task<string> stringDataTask = GetStringDataAsync();
DoSomeOtherSynchronousWork();
string data = await stringDataTask;
await dataTask;
```

Using this second method, the application can execute some synchronous work while the two `async` methods continue to run on background threads. Once the synchronous work is complete, the application will await the two `async` methods.

Let's put our `async` knowledge to work in a more realistic sample project. In this example, we will create a new Windows client application with **WPF** that loads data from two `async` methods. We will simulate slow service calls to fetch the data in these methods by injecting non-blocking delays with `Task.Delay`. Each method will take several seconds to return its data, but the UI will remain responsive to user input:

1. Start by creating a new WPF project in Visual Studio. Name the project `AwaitWithWpf`.

2. Add two new classes to the project named `Order` and `MainViewModel`. Your solution should now look something like this:

Figure 4.1 – The AwaitWithWpf solution in Visual Studio

3. Next, open **NuGet Package Manager**, search for MVVM Toolkit on the **Browse** tab, and
    add the latest stable version of the Microsoft.Toolkit.Mvvm package to your project:

 Microsoft.Toolkit.Mvvm by Microsoft.Toolkit, **214K** downloads   7.1.2
This package includes a .NET Standard MVVM library with helpers such as:
   - ObservableObject: a base class for objects implementing the INotifyPropertyChanged interface.

Figure 4.2 – Adding the Microsoft.Toolkit.Mvvm package to the project

We will be using the **MVVM Toolkit** to add **Model-View-ViewModel (MVVM)**
functionality to our MainViewModel class.

> **Note**
>
> The MVVM Toolkit is an open source MVVM library that is part of the **Windows Community
> Toolkit** maintained by Microsoft. If you are unfamiliar with the MVVM pattern or the MVVM
> Toolkit, you can read more about them on Microsoft Docs: https://docs.microsoft.
> com/windows/communitytoolkit/mvvm/introduction.

4. Now, open the Order class and add the following implementation:

```
public class Order
{
    public int OrderId { get; set; }
    public string? CustomerName { get; set; }
    public bool IsArchived { get; set; }
}
```

This will provide a few properties to display for each order when the order list is populated
on MainWindow.

5. Now we will start to build the MainViewModel implementation. The first step is to add a
    list of orders to bind to the UI and a command to execute when we want to load the orders:

```
public class MainViewModel : ObservableObject
{
    private ObservableCollection<Order> _orders =
        new();
    public MainViewModel()
    {
        LoadOrderDataCommand = new AsyncRelayCommand
            (LoadOrderDataAsync);
    }
```

```
public ICommand LoadOrderDataCommand { get; set; }
public ObservableCollection<Order> Orders
{
    get { return _orders; }
    set
    {
        SetProperty(ref _orders, value);
    }
}
private async Task LoadOrderDataAsync()
{
    // TODO - Add code to load orders
}
}
```

Let's review a few of the properties of the `MainViewModel` class before moving on to the next step:

- The `MainViewModel` class inherits from the `ObservableObject` type provided by the MVVM Toolkit.

- This base class implements the `INotifyPropertyChanged` interface, which is used by WPF data binding to notify the UI when data-bound property values change.

- The `Orders` property will provide the list of orders to the UI through WPF data binding. Calling `SetProperty` on the `ObservableObject` base sets the value of the `_orders` backing variable and triggers a property change notification.

- The `LoadOrderDataCommand` property will be executed by a button on `MainWindow`. In the constructor, the property is being initialized as a new `AsyncRelayCommand` that calls `LoadOrderDataAsync` when the command is invoked by the UI.

6. Don't forget to add the necessary `using` statements to the class:

```
using Microsoft.Toolkit.Mvvm.ComponentModel;
using Microsoft.Toolkit.Mvvm.Input;
using System.Collections.Generic;
using System.Collections.ObjectModel;
using System.Threading.Tasks;
using System.Windows.Input;
```

7. Next, let's create two `async` methods to load order data. One will create current orders and the other will create a list of archived orders. These are differentiated by the `IsArchived`

property on the Order class. Each method uses Task.Delay to simulate a service call across a slow internet or network connection:

```
private async Task<List<Order>> GetCurrentOrders
    Async()
{
    var orders = new List<Order>();
    await Task.Delay(4000);
    orders.Add(new Order { OrderId = 55, CustomerName
        = "Tony", IsArchived = false });
    orders.Add(new Order { OrderId = 56, CustomerName
        = "Peggy", IsArchived = false });
    orders.Add(new Order { OrderId = 60, CustomerName
        = "Carol", IsArchived = false });
    orders.Add(new Order { OrderId = 62, CustomerName
        = "Bruce", IsArchived = false });
    return orders;
}
private async Task<List<Order>> GetArchivedOrders
    Async()
{
    var orders = new List<Order>();
    await Task.Delay(5000);
    orders.Add(new Order { OrderId = 3, CustomerName =
        "Howard", IsArchived = true });
    orders.Add(new Order { OrderId = 18, CustomerName
        = "Steve", IsArchived = true });
    orders.Add(new Order { OrderId = 19, CustomerName
        = "Peter", IsArchived = true });
    orders.Add(new Order { OrderId = 21, CustomerName
        = "Mary", IsArchived = true });
    orders.Add(new Order { OrderId = 25, CustomerName
        = "Gwen", IsArchived = true });
    orders.Add(new Order { OrderId = 34, CustomerName
        = "Harry", IsArchived = true });
    orders.Add(new Order { OrderId = 36, CustomerName
        = "Bob", IsArchived = true });
```

```
        orders.Add(new Order { OrderId = 49, CustomerName
            = "Bob", IsArchived = true });
        return orders;
    }
```

8. Now we need to create a synchronous `ProcessOrders` method that combines the two lists of orders and updates the `Orders` property with the full dataset:

```
private void ProcessOrders(List<Order> currentOrders,
    List<Order> archivedOrders)
{
    List<Order> allOrders = new(currentOrders);
    allOrders.AddRange(archivedOrders);
    Orders = new ObservableCollection<Order>
        (allOrders);
}
```

9. The final step in building the `MainViewModel` class is the most important. Add the following implementation to the `LoadOrderDataAsync` method:

```
private async Task LoadOrderDataAsync()
{
    Task<List<Order>> currentOrdersTask =
        GetCurrentOrdersAsync();
    Task<List<Order>> archivedOrdersTask =
        GetArchivedOrdersAsync();
    List<Order>[] results = await Task.WhenAll(new
        Task<List<Order>>[] {
        currentOrdersTask, archivedOrdersTask
    }).ConfigureAwait(false);
    ProcessOrders(results[0], results[1]);
}
```

This method calls `GetCurrentOrdersAsync` and `GetArchivedOrdersAsync` and captures each in a `Task<List<Order>>` variable. You could simply await each call and store the returned orders in `List<Order>` variables. However, that would mean the second method would not start executing until the first one completed. By awaiting `Task.WhenAll` instead, the methods can execute in parallel on background threads.

If your methods all return the same data type, you can capture the results of `Task.WhenAll` in an array of the return type. In our case, we are receiving the two lists of orders in an array of `List<Order>` and passing the two array values to `ProcessOrders`.

10. Now, let's move on to the `MainWindow.xaml.cs` code-behind file. Add the following code to set `DataContext` of `MainWindow` in the constructor after the call to `InitializeComponent`:

```
public MainWindow()
{
    InitializeComponent();
    var vm = new MainViewModel();
    DataContext = vm;
}
```

`DataContext` is the source for all `Binding` references in the XAML for `MainWindow`. We will create the XAML for our UI in the next step.

11. The last file to update is `MainWindow.xaml`. Open the XAML file and start by adding two rows to `Grid`. The first row will contain another `Grid` containing `Button` and `TextBox`. The second row will contain `ListView` to display the list of orders. We'll create a template for the orders in a moment:

```
<Grid>
    <Grid.RowDefinitions>
        <RowDefinition Height="Auto"/>
        <RowDefinition Height="*"/>
    </Grid.RowDefinitions>
    <Grid Grid.Row="0" Margin="4">
        <Grid.ColumnDefinitions>
            <ColumnDefinition Width="Auto"/>
            <ColumnDefinition Width="*"/>
        </Grid.ColumnDefinitions>
        <Button Content="Load Data" Grid.Column="0"
            Margin="2" Width="200"
            Command="{Binding Path=LoadOrderData
                Command}"/>
        <TextBox Grid.Column="1" Margin="2"/>
    </Grid>
    <ListView Grid.Row="1" ItemsSource="{Binding
        Path=Orders}" Margin="4">
```

```
    </ListView>
</Grid>
```

I have highlighted the two data binding instances in the XAML markup. The Command of Button is bound to the LoadOrderDataCommand property, and ItemsSource of ListView is bound to the Orders property. Setting ItemsSource will make the properties of the Order class available to the members of ListView.ItemTemplate.

12. Let's add ItemTemplate to ListView next. Defining DataTemplate within ItemTemplate defines the structure of each item within ListView:

```
<ListView Grid.Row="1" ItemsSource="{Binding
    Path=Orders}" Margin="4">
    <ListView.ItemTemplate>
        <DataTemplate>
            <StackPanel Margin="2">
                <StackPanel Orientation="Horizontal">
                    <TextBlock Text="Order Id:"
                            Margin="2,2,0,2"
                            Width="100"/>
                    <TextBox IsReadOnly="True"
                            Width="200"
                            Text="{Binding
                            Path=OrderId}" Margin="2"/>
                </StackPanel>
                <StackPanel Orientation="Horizontal">
                    <TextBlock Text="Customer:"
                            Margin="2,2,0,2"
                            Width="100"/>
                    <TextBox IsReadOnly="True"
                            Width="200"
                            Text="{Binding
                            Path=CustomerName}"
                            Margin="2"/>
                </StackPanel>
                <StackPanel Orientation="Horizontal">
                    <TextBlock Text="Archived:"
                            Margin="2,2,0,2"
                            Width="100"/>
```

```
                           <TextBox IsReadOnly="True"
                                    Width="200"
                                    Text="{Binding
                                    Path=IsArchived}"
                                    Margin="2"/>
                       </StackPanel>
                    </StackPanel>
                </DataTemplate>
            </ListView.ItemTemplate>
        </ListView>
```

Each `Order` instance will render as a `StackPanel` containing three horizontally aligned `StackPanel` elements, displaying labels and values for the `OrderId`, `CustomerName`, and `IsArchived` data-bound properties.

13. We're ready to run the application and see how things work. After the program starts, click the **Load Data** button. It will take about 5 seconds to load the data to `ListView`. While you wait, try typing some text into the box to the right of the **Load Data** button. You can see that the UI remains responsive to user input thanks to `async`/`await` and the `Task.WhenAll` method. Once the data has finished loading, you should see a list of twelve orders in the scrollable list:

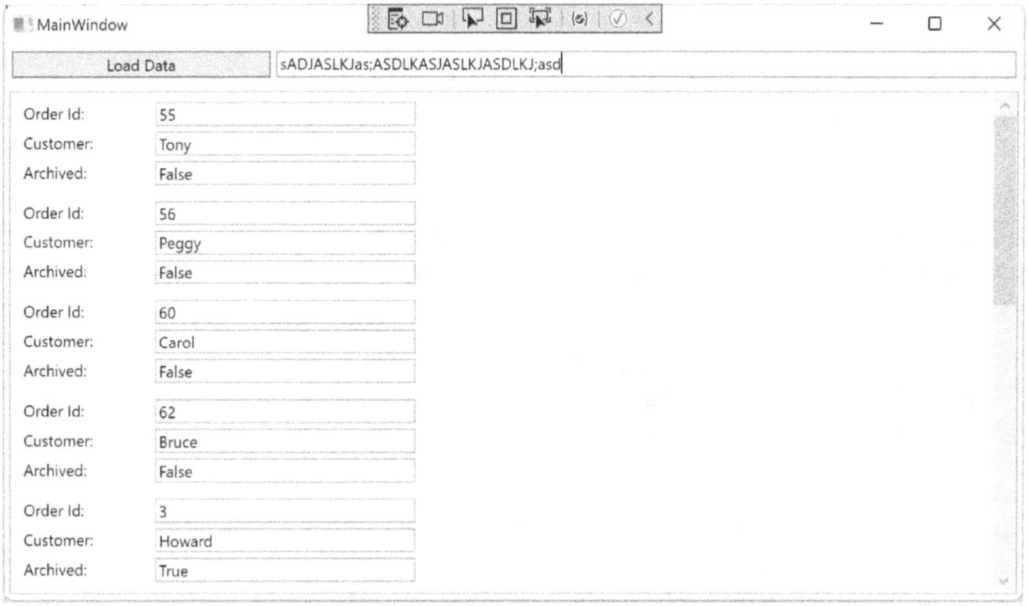

Figure 4.2 – Viewing a list of orders in the AsyncWithWpf application

In a real production application, the implementations of the two `async` methods would be replaced by service calls to fetch data from a database or web services. Regardless of how long it takes to return and populate the data, other parts of the UI will remain responsive to user input. One change you would want to make is adding an indicator to the UI to inform the user that data is being loaded. You should also disable the **Load Data** button while the data load process is active to prevent multiple calls to `LoadOrderDataAsync`.

The example illustrates the benefits of using `async` and `await` in a Windows application. These `async` calls are using `ThreadPool` within the TPL. Let's look at some other ways to leverage `ThreadPool` in a Windows application.

Using the thread pool

There are other ways to use `ThreadPool` threads in a .NET application. Let's discuss a situation where you want to accomplish the same result that was achieved with `async` and `await` in the previous example, but the methods to fetch the order data are not marked as `async`. One option is to update the methods to be `async`. If that code is not within your control to change, you have some other options available.

The `ThreadPool` class has a method called `QueueUserWorkItem`. This method accepts a method to call and queues it for execution on the thread pool. We could use it with our project like this:

```
ThreadPool.QueueUserWorkItem(GetCurrentOrders);
```

There are a few problems with using this method. The primary issue is that there is no return value to get the list of orders from the method call. You could work around this issue with some wrapper methods that update a shared thread-safe collection such as the `BlockingCollection`. That isn't a great design, and there is a better option.

The `QueueUserWorkItem` method was more commonly used before the introduction of the TPL. In today's task-based world, you can use `Task.Run` to execute a synchronous method as `async`. Let's update our WPF project to use `Task.Run`:

1. The only file that needs to be modified to use `Task.Run` is `MainViewModel`. Start by updating `GetCurrentOrdersAsync` and `GetArchivedOrdersAsync` to no longer be `async` methods. They should also be renamed as `GetCurrentOrders` and `GetArchivedOrders` so consumers are aware that they are not `async` methods:

    ```
    private List<Order> GetCurrentOrders()
    {
        var orders = new List<Order>();
        Thread.Sleep(4000);
        orders.Add(new Order { OrderId = 55, CustomerName
    ```

```
                = "Tony", IsArchived = false });
        orders.Add(new Order { OrderId = 56, CustomerName
                = "Peggy", IsArchived = false });
        orders.Add(new Order { OrderId = 60, CustomerName
                = "Carol", IsArchived = false });
        orders.Add(new Order { OrderId = 62, CustomerName
                = "Bruce", IsArchived = false });
        return orders;
    }
    private List<Order> GetArchivedOrders()
    {
        var orders = new List<Order>();
        Thread.Sleep(5000);

        orders.Add(new Order { OrderId = 3, CustomerName =
            "Howard", IsArchived = true });
        orders.Add(new Order { OrderId = 18, CustomerName
                = "Steve", IsArchived = true });
        orders.Add(new Order { OrderId = 19, CustomerName
                = "Peter", IsArchived = true });
        orders.Add(new Order { OrderId = 21, CustomerName
                = "Mary", IsArchived = true });
        orders.Add(new Order { OrderId = 25, CustomerName
                = "Gwen", IsArchived = true });
        orders.Add(new Order { OrderId = 34, CustomerName
                = "Harry", IsArchived = true });
        orders.Add(new Order { OrderId = 36, CustomerName
                = "Bob", IsArchived = true });
        orders.Add(new Order { OrderId = 49, CustomerName
                = "Bob", IsArchived = true });
        return orders;
    }
```

The changes are minimal, and I have highlighted them in the preceding source code. The async modifier has been removed from the method declarations, the methods have been renamed and they no longer return tasks, and Task.Delay in each method has been updated to Thread.Sleep.

2. Next, we will update the LoadOrderDataAsync method to call the synchronous methods with Task.Run:

    ```
    private async Task LoadOrderDataAsync()
    {
        Task<List<Order>> currentOrdersTask =
            Task.Run(GetCurrentOrders);
        Task<List<Order>> archivedOrdersTask =
            Task.Run(GetArchivedOrders);
        List<Order>[] results = await Task.WhenAll(new
            Task<List<Order>>[] {
            currentOrdersTask, archivedOrdersTask
        }).ConfigureAwait(false);
        ProcessOrders(results[0], results[1]);
    }
    ```

    No other changes are necessary. Task.Run will return the same Task<List<Order>> type, which can still be used with Task.WhenAll to wait for their completion.

3. Run the program, and it should work exactly as it did before. The UI remains responsive while the order data is loading.

This is an excellent way to start incorporating async and await into existing code, but always use caution when adding threading to your applications. In this application, the two methods being called do not access any shared data. So, there was no need to think about thread safety. If these methods were updating a private collection of orders, you would need to introduce a locking mechanism or use a thread-safe collection for the orders.

Before we move on to a discussion of the UI thread, there is one other Task method to discuss. The Task.Factory.StartNew method is similar in use to Task.Run. In fact, you can use them in the same way. This code uses Task.Run to get a Task with the current orders:

```
Task<List<Order>> currentOrdersTask = Task.Run
    (GetCurrentOrders);
```

This code does the same thing with Task.Factory.StartNew:

```
Task<List<Order>> currentOrdersTask = Task.Factory.StartNew
    (GetCurrentOrders);
```

In this case, you should use Task.Run. It is a newer method and is simply a shortcut meant to simplify the most common use cases. The Task.Factory.StartNew method has some additional overloads for specific uses. This example uses `StartNew` to call `GetCurrentOrders` with some optional parameters:

```
Task<List<Order>> currentOrdersTask =
    Task.Factory.StartNew(GetCurrentOrders,
    CancellationToken.None,
    TaskCreationOptions.AttachedToParent,
    TaskScheduler.Default);
```

The interesting option we have provided here is `TaskCreationOptions.AttachedToParent`. What this does is it links the task completion of the calling method to that of the child, `GetCurrentOrders`. The default behavior is for their completions to be unlinked. For a complete list of available overloads and their uses, you can review Microsoft Docs here: `https://docs.microsoft.com/dotnet/api/system.threading.tasks.taskfactory.startnew`.

> **Note**
>
> *Stephen Toub* of the .NET team has a blog post where he discusses `Task.Run` versus `Task.Factory.StartNew` and why you might want to choose each option. You can read his post on the *.NET Parallel Programming* blog here: `https://devblogs.microsoft.com/pfxteam/task-run-vs-task-factory-startnew/`.

Now, let's move on to discuss when you will need to write code to explicitly update the UI thread from a background thread.

Updating the UI thread without exceptions

When working with managed threading in .NET applications, there are many pitfalls that developers must learn to avoid. One of the common mistakes developers make is writing code that updates a UI control in a Windows application from a non-UI thread. This kind of error will not be detected by the compiler. Developers will receive a runtime error indicating that a control created on the main thread cannot be modified on another thread.

So, how do you avoid these runtime errors? The best way to avoid them is by not updating UI controls from background threads at all. WPF helps avoid the problem with the MVVM pattern and data binding. Binding updates are automatically marshaled to the UI thread by .NET. You can safely update properties in your `ViewModel` classes from a background thread without causing errors at runtime.

If you are updating UI controls directly in your code, either in a WinForms application or in the code-behind file of a WPF control, you can use an `Invoke` call to *push* the execution to the main thread. The implementation is slightly different between WinForms and WPF. Let's start with a WPF example. If you have a method performing some work on a background thread, and it needs to update the `Text` property of a `TextBox` on a WPF window, you could wrap the code in an action:

```
Application.Current.Dispatcher.Invoke(new Action(() => {
    usernameTextBox.Text = "John Doe";
}));
```

`Dispatcher.Invoke` will push the execution to the main thread. Keep in mind that if the main thread is busy with other work, your background thread will wait here for this action to complete. If your background worker wants to fire and forget this action, you can use `Dispatcher.BeginInvoke` instead.

Let's assume we want to update `usernameTextBox`, but this time, we are working with a WinForms project. The same invocation can be accomplished by using `Form` or `UserControl` executing the code. This example is a WinForms application with two buttons. Clicking one button will call the `UpdateUsername` method. The other button will call `Task.Run(UpdateUsername)`, putting it on a background thread. To determine whether `Invoke` is needed to access the main thread, you check the Boolean `InvokeRequired` read-only property. It may not be required if the thread pool chose to run `Task` on the main thread:

```
public partial class Form1 : Form
{
    public Form1()
    {
        InitializeComponent();
    }
    private void btnRunInBackground_Click(object sender,
        EventArgs e)
    {
        Task.Run(UpdateUsername);
    }
    private void btnRunOnMainThread_Click(object sender,
        EventArgs e)
    {
        UpdateUsername();
    }
    private void UpdateUsername()
```

```
    {
        var updateAction = new Action(() =>
        {
            usernameTextBox.Text = "John Doe";
        });
        if (this.InvokeRequired)
        {
            this.Invoke(updateAction);
        }
        else
        {
            updateAction();
        }
    }
}
```

The usernameTextBox will display the name **John Doe** successfully regardless of which button is clicked:

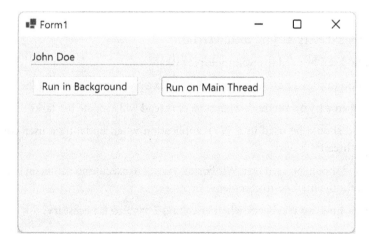

Figure 4.3 – Updating a control on a WinForms form

Like WPF, WinForms has a BeginInvoke method if the background code does not need to wait for the main thread update to complete. BeginInvoke can also accept an EndInvoke delegate that will receive a callback when the main thread invocation has completed.

This section provided a great start on using .NET managed threading in your Windows client applications. Let's finish up with a summary of what we learned in this chapter.

Summary

In this chapter, we learned some useful techniques for improving client application performance. We started by exploring some different uses of `async` and `await` in the ViewModel of a WPF application. In that project, we saw that awaiting `Task.WhenAll` does not block the main thread, which keeps the UI responsive to user input. We discussed how `Task.Run` and `Task.Factory.StartNew` can be used to call synchronous code from asynchronous code, making it easier to introduce managed threading to existing applications. We finished up the chapter by learning some techniques to update the UI thread from other threads without causing exceptions at runtime.

You should be feeling more comfortable using `async`, `await`, and the TPL in your code after reading this chapter. Try taking what you have learned here and start adding some `async` code to your own client applications. For additional reading on `async` and `await`, you can check out this C# article on Microsoft Docs: `https://docs.microsoft.com/dotnet/csharp/async`.

In the next chapter, we will dive even deeper into using `async`, `await`, and the TPL. We will take some of the concepts from this chapter and expand on them while introducing some best practices.

Questions

1.   What type should every `async` method return?

2.   Which method can be used to await multiple tasks?

3.   Which method to start a new task accepts `TaskDispatcher` as one of the parameters?

4.   When calling an `async` method, what type of thread will execute the task?

5.   What method should be used in a WPF application when updating a user control from a background thread?

6.   Which method should be used on a WinForms control to execute an action on the main thread but not wait for the method to complete?

7.   In WinForms, how can you check whether calling `Invoke` is necessary?

Part 2: Parallel Programming and Concurrency with C#

It's time to dive deeper into the modern methods of parallel programming and concurrency with C# and .NET 6. This part will explore some of the most common real-world practices employed today.

This part contains the following chapters:

- *Chapter 5, Asynchronous Programming with C#*
- *Chapter 6, Parallel Programming Concepts*
- *Chapter 7, Task Parallel Library (TPL) and Dataflow*
- *Chapter 8, Parallel Data Structures and Parallel LINQ (PLINQ)*
- *Chapter 9, Working with Concurrent Collections in .NET*

5

Asynchronous Programming with C#

The .NET **task asynchronous programming** (**TAP**) model, which uses the `async` and `await` keywords, was introduced in .NET Framework 4.5. The C# language's support for these keywords was released at the same time in C# 5. Now, a decade later, the TAP model is an integral part of most .NET developers' toolsets.

This chapter will explain asynchronous programming in C#, explore how to use `Task` objects, and delve into best practices of using `async` and `await` for **I/O-bound** and **CPU-bound** scenarios with .NET.

In this chapter, you will learn about the following:

- More about asynchronous programming in .NET
- Working with Task objects
- Interop with synchronous code
- Working with multiple background tasks
- Asynchronous programming best practices

By the end of this chapter, you will have a deeper understanding of asynchronous programming and should feel confident enough to add advanced async features to your team's projects.

Technical requirements

In this chapter, we will be using the .NET **command-line interface (CLI)** and Visual Studio Code to build and run the sample projects. To follow along with the examples, the following software is recommended:

- Visual Studio Code version 1.65 or later

- .NET 6 or later

While these are recommended, if you have .NET 6 installed, you can use your preferred editor. For example, Visual Studio 2022 version 17.0 or later if you are using Windows 10 or 11, Visual Studio 2022 for Mac on macOS 10.13 or later, or JetBrains Rider will work just as well.

All the code examples for this chapter can be found on GitHub at `https://github.com/PacktPublishing/Parallel-Programming-and-Concurrency-with-C-sharp-10-and-.NET-6/tree/main/chapter05`.

Let's get started by working our way through some examples that use the TAP model with `async` and `await`.

More about asynchronous programming in .NET

There are two types of scenarios where async code is usually introduced:

- **I/O-bound operations**: These involve resources fetched from the network or disk.

- **CPU-bound operations**: These are in-memory, CPU-intensive operations.

In this section, we will create some real-world examples that use `async` and `await` for each type of operation. Whether you are waiting for an external process to complete or performing CPU-intensive operations within your application, you can leverage asynchronous code to improve your application's performance.

Let's start by looking at some examples of I/O-bound operations.

I/O-bound operations

When you are working with I/O-bound code that is constrained by file or network operations, your code should use `async` and `await` to wait for the operations to complete.

The .NET methods to perform network and file I/O are asynchronous, so the use of Task.Run will not be necessary:

- **Example 1**: Let's look at an example of an async method that reads the contents of a text file with the ReadToEndAsync method, splits the text where Environment.NewLine characters are found, and returns the data as a List<string> instance. Each line of text from the file is an item in the list:

```
public async Task<List<string>> GetDataAsync
    (string filePath)
{
    using var file = File.OpenText(filePath);
    var data = await file.ReadToEndAsync();
    return data.Split(new[] { Environment.NewLine },
        StringSplitOptions.RemoveEmptyEntries)
            .ToList();
}
```

- **Example 2**: Another example of I/O-bound operations is a **file download**. We will take the concept from the previous example, but this time the file to be split and returned is hosted on a web server on the network. We will use the HttpClient class to download a file from the provided URL with the await keyword before splitting and returning the lines of text in a list:

```
public async Task<List<string>> GetOnlineDataAsync
    (string url)
{
    var httpClient = new HttpClient();
    var data = await httpClient.GetStringAsync(url);
    return data.Split(new[] { Environment.NewLine },
        StringSplitOptions.RemoveEmptyEntries)
            .ToList();
}
```

Those are some common I/O-bound operations, but what is a CPU-bound operation and how does it differ?

CPU-bound operations

In this case, your application is not waiting for an external process to complete. The application itself is performing a CPU-intensive operation that takes time to complete, and you want the application to remain responsive until the operation has finished.

In this example, we have a method that accepts a `List<string>` instance where each item in the list contains an XML representation of this `JournalEntry` class:

```csharp
[Serializable]
public class JournalEntry
{
    public string Title { get; set; }
    public string Description { get; set; }
    public DateTime EntryDate { get; set; }
    public string EntryText { get; set; }
}
```

Let's assume that `EntryText` can be extremely large because some users who write in the journal application will add dozens of pages of text to a single entry. Each entry is stored in a database as XML and the application that loads the entries has a `DeserializeEntries` method to deserialize each XML string and return the data as a `List<JournalEntry>` instance:

```csharp
private List<JournalEntry> DeserializeEntries(List<string>
    journalData)
{
    var deserializedEntries = new List<JournalEntry>();
    var serializer = new XmlSerializer(typeof
        (JournalEntry));
    foreach (var xmlEntry in journalData)
    {
        if (xmlEntry == null) continue;
        using var reader = new StringReader(xmlEntry);
        var entry = (JournalEntry)serializer.Deserialize
            (reader)!;
        if (entry == null) continue;
        deserializedEntries.Add(entry);
    }
    return deserializedEntries;
}
```

After months of adding journal entries, users are complaining about the time it takes to load the existing entries. They would like to start creating a new entry while the data is loading.

Luckily, using asynchronous .NET code can keep an application's user interface responsive while waiting for a long-running process to complete. The thread is free to perform other work until the non-blocking call is completed. By adding an async method named DeserializeJournalDataAsync that calls the existing method with an awaited Task.Run method, the client code can remain responsive while users create new journal entries:

```
public async Task<List<JournalEntry>>
    DeserializeJournalDataAsync(List<string> journalData)
{
    return await Task.Run(() => DeserializeEntries
        (journalData));
}
```

If you're working with serialized data in JSON format instead of XML, the synchronous and asynchronous methods of deserialization are very similar. This is because .NET provides both the Deserialize and DeserializeAsync methods in the System.Text.Json.JsonSerializer class. Here are both methods with their differences highlighted:

```
public List<JournalEntry> DeserialzeJsonEntries
    (List<string> journalData)
{
    var deserializedEntries = new List<JournalEntry>();
    foreach (var jsonEntry in journalData)
    {
        if (string.IsNullOrWhiteSpace(jsonEntry)) continue;
        deserializedEntries.Add(JsonSerializer.Deserialize
            <JournalEntry>(jsonEntry)!);
    }
    return deserializedEntries;
}
public async Task<List<JournalEntry>> Deserialize
    JsonEntriesAsync(List<string> journalData)
{
    var deserializedEntries = new List<JournalEntry>();
    foreach (var jsonEntry in journalData)
    {
```

```
        if (string.IsNullOrWhiteSpace(jsonEntry)) continue;
        using var stream = new MemoryStream(Encoding
            .Unicode.GetBytes(jsonEntry));
        deserializedEntries.Add((await JsonSerializer
            .DeserializeAsync<JournalEntry>(stream))!);
    }
    return deserializedEntries;
}
```

The Deserialize method accepts string, but DeserializeAsync does not. Instead, we must create a MemoryStream instance from the jsonEntry string to pass to DeserializeAsync. Other than that, only the return types of the methods differ.

Let's wrap up this section by looking at one more method for handling JSON deserialization of a list of journal entries. In this example, the method that deserializes the data only processes a single JSON entry. A parent method named GetJournalEntriesAsync uses a LINQ Select operator to call DeserializeJsonEntryAsync for each string in the list and stores an IEnumerable<Task<JournalEntry>> instance in a getJournalTasks variable:

```
public async Task<List<JournalEntry>>
    GetJournalEntriesAsync(List<string> journalData)
{
    var journalTasks = journalData.Select(entry =>
        DeserializeJsonEntryAsync(entry));
    return (await Task.WhenAll(journalTasks)).ToList();
}
private async Task<JournalEntry> DeserializeJsonEntryAsync
    (string jsonEntry)
{
    if (string.IsNullOrWhiteSpace(jsonEntry)) return new
        JournalEntry();
    using var stream = new MemoryStream
        (Encoding.Unicode.GetBytes(jsonEntry));
    return (await JsonSerializer.DeserializeAsync
        <JournalEntry>(stream))!;
}
```

The highlighted code awaits all the Task objects in journalTasks, returning the results of every call as an array of JournalEntry objects. You can either declare GetJournalEntriesAsync with a return type of Task<JournalEntry[]> or use ToList, as we have in this sample, to return Task<List<JournalEntry>>. You can see how LINQ streamlines your code when it is necessary to iterate over a list of items and make an async call with each item.

You have seen some different ways to use async and await in your code for both I/O-bound and CPU-bound operations.

Next, we will discuss how **nested async methods** are chained and how to start the top level of that chain.

Nested async methods

When it comes to using async methods, it is important to use await when you want to preserve the order of execution. It is also important to preserve that chain of awaited calls to the entry point for the current thread.

For example, if your application is a console application, the primary entry point is the Main method in Program.cs. If you cannot make this Main method async, then none of the method calls beneath Main are made with the await keyword. That is the reason why .NET now supports async Main methods. Now, when you create a new console application with .NET 6, it has an async Main method by default.

If the entry point for execution is an event handler, you should mark the event handler method as async. This is the only time you will see async methods with a void return type:

```
private async void saveButton_Click(object sender,
    EventArgs e)
{
    await SaveData();
}
```

Let's look at an example of the right way to chain multiple nested async methods in a console application:

1. Start by creating a new console application. Inside a folder named AsyncSamples, run the following command:

    ```
    dotnet new console -framework net6.0
    ```

2. When the process completes, open the new AsyncSamples.csproj in Visual Studio Code or your editor of choice.

3. Add a new class to the project named `TaskSample`

4. Add the following code to the `TaskSample` class:

```csharp
public async Task DoThingsAsync()
{
    Console.WriteLine($"Doing things in
        {nameof(DoThingsAsync)}");
    await DoFirstThingAsync();
    await DoSecondThingAsync();
    Console.WriteLine($"Did things in
        {nameof(DoThingsAsync)}");
}
private async Task DoFirstThingAsync()
{
    Console.WriteLine($"Doing something in
        {nameof(DoFirstThingAsync)}");
    await DoAnotherThingAsync();
    Console.WriteLine($"Did something in
        {nameof(DoFirstThingAsync)}");
}
private async Task DoSecondThingAsync()
{
    Console.WriteLine($"Doing something in
        {nameof(DoSecondThingAsync)}");
    await Task.Delay(500);
    Console.WriteLine($"Did something in
        {nameof(DoSecondThingAsync)}");
}
private async Task DoAnotherThingAsync()
{
    Console.WriteLine($"Doing something in
        {nameof(DoAnotherThingAsync)}");
```

```
        await Task.Delay(1500);
        Console.WriteLine($"Did something in
            {nameof(DoAnotherThingAsync)}");
    }
```

5. Now open `Program.cs` and add some code to call `DoThingsAsync`:

```
    using AsyncSamples;
    Console.WriteLine("Start processing"…");
    var taskSample = new TaskSample();
    await taskSample.DoThingsAsync();
    Console.WriteLi"e("Done processing"..");
```

Let's illustrate the order and hierarchy of the methods being called by our project. The `Main` method calls `DoThingsAsync`, which in turn calls `DoFirstThingAsync` and `DoSecondThingAsync`. Finally, within `DoFirstThingAsync`, `DoAnotherThingAsync` is called. When each of these `async` methods is called with the `await` operator, the order of operations is predictable:

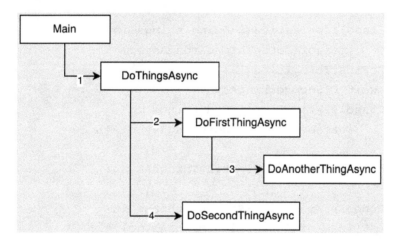

Figure 5.1: The order of operations for awaited methods

6. Run the program and examine the order of the console output. Everything should be executing in the expected order:

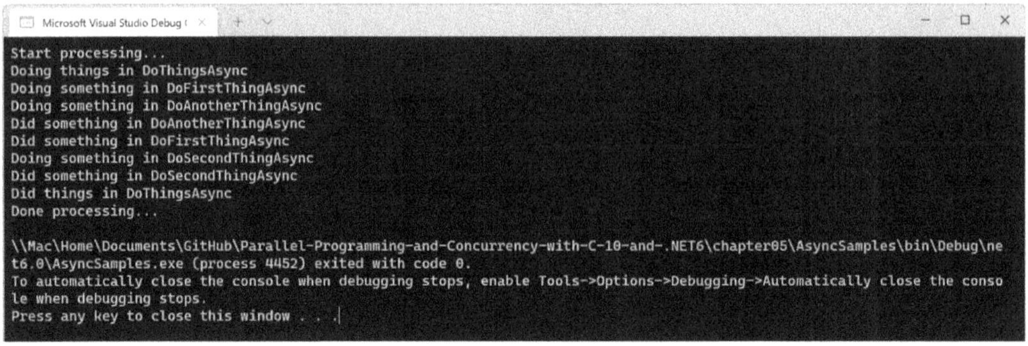

Figure 5.2: Examining the output of the AsyncSamples console application

7. Next, we will add two additional methods to the `TaskSample` class:

```
public async Task DoingThingsWrongAsync()
{
    Console.WriteLine($"Doing things in
        {nameof(DoingThingsWrongAsync)}");
    DoFirstThingAsync();
    await DoSecondThingAsync();
    Console.WriteLine($"Did things in
        {nameof(DoingThingsWrongAsync)}");
}
public async Task DoBlockingThingsAsync()
{
    Console.WriteLine($"Doing things in
        {nameof(DoBlockingThingsAsync)}");
    DoFirstThingAsync().Wait();
    await DoSecondThingAsync();
    Console.WriteLine($"Did things in
        {nameof(DoBlockingThingsAsync)}");
}
```

The `DoingThingsWrongAsync` method has removed the `await` from the call to `DoFirstThingAsync`. So, the execution of `DoSecondThingAsync` will begin before `DoFirstThingAsync` has been completed. That might be OK if none of the subsequent code relies on the processing that happens within `DoFirstThingAsync`. However, any unhandled exceptions inside a method that is not awaited will not automatically bubble up to the `calling` method. The `Task` instance for the call will have a `Status` value of `Faulted`, the `IsFaulted` property will be `true`, and the `Exception` property will contain the unhandled exception information.

In the preceding case, any unhandled exceptions in `DoFirstThingAsync` will go undetected. If you have a case where you are not awaiting a `Task` instance, be sure to monitor the status of the `Task` instance in case of exceptions. This is one of the reasons why you should never have an `async void` method. It does not return a `Task` instance to be awaited.

The `DoBlockingThings` method will maintain the correct order of operations, but by calling `DoFirstThingAsync().Wait()` instead of awaiting the call, the thread executing `DoBlockingThings` will be blocked. It will wait for the call to `DoFirstThingAsync` to complete instead of being free to pick up other work until the long-running async method completes. Using blocking calls such as `Wait()` or `Result` can quickly deplete the available threads in `ThreadPool`.

8. Update `Program.cs` to call all three of the public `TaskSample` methods:

    ```
    using AsyncSamples;
    Console.WriteLine("Start processing...");
    var taskSample = new TaskSample();
    await taskSample.DoThingsAsync();
    Console.WriteLine("Continue processing...");
    await taskSample.DoingThingsWrongAsync();
    Console.WriteLine("Continue processing...");
    await taskSample.DoBlockingThingsAsync();
    Console.WriteLine("Done processing...");
    ```

9. Now run the program and examine the console output to see how it is impacted by omitting `await` inside `DoingThingsWrongAsync`:

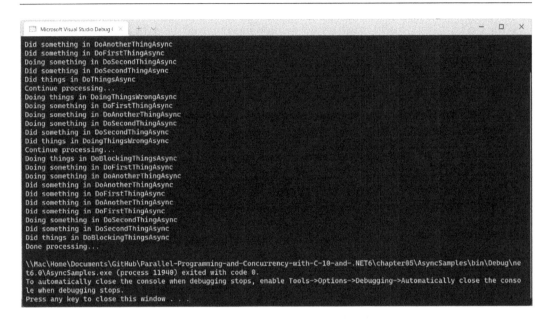

Figure 5.3: Console output when calling all the TaskSample methods

The output may differ a little each time, depending on how the `ThreadPool` threads are allocated. In this case, the second call to `DoFirstThingAsync` remains incomplete until the third call to that same method starts. Even though `Program.cs` awaits its call to `DoingThingsWrongAsync`, the code inside of that method was still executing after the next call to `DoBlockingThingsAsync` was invoked.

Things can get very unpredictable when async tasks are not awaited. You should always await a task unless you have a good reason not to do so. Next, let's explore some properties and methods available in the `Task` class.

Working with Task objects

Working directly with `Task` objects can be extremely useful when introducing threading to existing projects. As we saw in the previous section, it is important to update the entire call stack when introducing `async` and `await`. On a large code base, those changes could be extensive and would require quite a bit of regression testing.

You can instead use `Task` and `Task<TResult>` to wrap the existing methods that you want to run asynchronously. Both `Task` types represent the asynchronous work being done by a method or action. You use `Task` when a method would have otherwise returned void. Use `Task<TResult>` with methods that have a non-void return type.

Here are examples of two synchronous method signatures and their async equivalents:

```
public interface IAsyncExamples
{
    void ProcessOrders(List<Order> orders);
    Task ProcessOrdersAsync(List<Order> orders);
    List<Order> GetOrders(int customerId);
    Task<List<Order>> GetOrdersAsync(int customerId);
}
```

We have seen some examples of using `Task` objects in this chapter. Now it is time to explore additional properties, methods, and uses of these two types.

Exploring Task methods

To start, we will discover some commonly used `Task` methods in practical examples. Consider the ProcessOrders method that accepts a list of orders to be processed and submitted. The four `Task` methods used are as follows:

- `Task.Run`: Runs a method on a thread on the thread pool

- `Task.Factory.StartNew`: Runs a method on a thread on the thread pool, with `TaskCreationOptions` provided

- `processOrdersTask.ContinueWith`: When the `processOrdersTask` completes, it will execute the method provided on the same thread pool thread.

- `Task.WaitAll`: This method will block the current thread and wait for all tasks in the array.

These methods have been highlighted in the following code:

```
public void ProcessOrders(List<Order> orders, int
    customerId)
{
    Task<List<Order>> processOrdersTask = Task.Run(() =>
        PrepareOrders(orders));
    Task labelTask = Task.Factory.StartNew(() =>
        CreateLabels(orders), TaskCreationOptions
            .LongRunning);
    Task sendTask = processOrdersTask.ContinueWith(task =>
        SendOrders(task.Result));
```

```
    Task.WaitAll(new[] { labelTask, sendTask });
    SendConfirmation(customerId);
}
```

This is what is happening on each line of the preceding example:

1. `Task.Run` will create a new background thread and queue it on `ThreadPool`

2. `Task.Factory.StartNew` will also create a new background thread and queue it on `ThreadPool`. In addition, we are providing `TaskCreattionOptions.LongRunning` as a parameter of `StartNew` to indicate that creating additional threads is warranted because this task may take a while to complete. This will prevent delays for other tasks queued on `ThreadPool`.

3. `ContinueWith` will queue `SendOrders` on a `ThreadPool` thread, but the thread will not start until `processOrdersTask` has been completed.

4. `Task.WaitAll` is the synchronous equivalent of the `async` method, `Task.WhenAll`. It will block the current thread until `labelTask` and `sendTask` are complete.

5. Finally, `SendConfirmation` is called to notify the customer that their orders have been processed and sent.

Using tasks in this way can achieve the same result as an `async` method that awaits tasks to achieve parallel processing. The main difference is that the current thread will be blocked at *step 4* when `WaitAll` is called.

Another useful method we will explore next is `RunSynchronously`. This starts a task but executes it synchronously on the current thread. The asynchronous equivalent is to call `Start` on a task.

In this example, the `ProcessData` method accepts a parameter indicating whether the data must be processed on the UI thread. It is possible that some data processing requires interacting with the UI to present the user with some options or other feedback:

```
public void ProcessData(object data, bool uiRequired)
{
    Task processTask = new(() => DoDataProcessing(data));
    if (uiRequired)
    {
        // Run on current thread (UI thread assumed for
            example)
        processTask.RunSynchronously();
    }
    else
```

```
    {
        // Run on ThreadPool thread in background
        processTask.Start();
    }
}
```

Next, let's explore some of the properties of the `Task` and `Task<TResult>` classes.

Exploring Task properties

In this section, we will review the properties available on a `Task` object. Most of the properties are related to the status of a task, so we will start with the `Status` property. The `Status` property returns `TaskStatus`, which is an enumeration with eight possible values:

- `Created (0)`: The task has been created and initialized but has not been scheduled on `ThreadPool`.

- `WaitingForActivation (1)`: The task is waiting to be scheduled by .NET

- `WaitingToRun (2)`: The task has been scheduled but has not started executing yet

- `Running (3)`: The task is currently running.

- `WaitingForChildrenToComplete (4)`: The task has been completed but there are attached child tasks that are still running or waiting to run

- `RanToCompletion (5)`: The task successfully ran to completion

- `Canceled (6)`: The task was canceled and acknowledged the cancellation

- `Faulted (7)`: An unhandled exception was encountered while executing the task

The following properties of `Task` and `Task<TResult>` are shortcuts to check statuses:

- `IsCanceled`: Returns `true` if the task's `Status` is `Canceled`

- `IsCompleted`: Returns `true` if the task's `Status` is `RanToCompletion`, `Canceled`, or `Faulted`

- `IsCompletedSuccessfully`: Returns `true` if the task's `Status` is `RanToCompletion`

- `IsFaulted`: Returns `true` if the task's `Status` is `Faulted`

Using these properties can streamline status checks in your code. The remaining instance properties of the `Task` object follow:

- `AsyncState`: Returns the state that was provided when creating the task. If no state was provided, this property returns `null`

- **CreationOptions**: Returns the **CreationOptions** values that were provided when creating the task. If no options were provided, it defaults to **TaskCreationOptions.None**.

- **Exception**: Returns an **AggregateException** instance containing unhandled exceptions encountered while the task was running. **Wait** or **WaitAll** should be called in a **try/catch** block that handles the **AggregateException** type.

- **Id**: A system-assigned identifier for the task

Let's take a quick look at how to correctly catch an **AggregateException** instance and inspect the **Exception** property of the faulted task:

```
Task ordersTask = Task.Run(() => ProcessOrders(orders,
    123));
try
{
    ordersTask.Wait();
    Console.WriteLine($"ordersTask Status:
        {ordersTask.Status}");
}
catch (AggregateException)
{
    Console.WriteLine($"Exception in ordersTask! Error
        message: {ordersTask.Exception.Message}");
}
```

This code will write the status of the task to the console after completion. If an unhandled exception is encountered, the error message will be written to the console in the **catch** block.

Now that you're more familiar with the members of **Task** and **Task<TResult>**, let's discuss some use cases for calling synchronous code from async code and vice versa.

Interop with synchronous code

When working with existing projects and introducing async code to the system, there will be points where synchronous and asynchronous code intersect. We have already seen some examples of how to handle this interop in this chapter. In this section, we will focus on that interop in both directions: sync calling async and async calling sync.

We will create a sample project with classes containing synchronous methods representing legacy code and another set of classes with modern **async** methods.

Let's start by discussing how to consume **async** methods in your legacy synchronous code.

Executing async from synchronous methods

In this example, we will be working with a .NET console application that gets a patient and their list of medications. The application will call a synchronous GetPatientAndMedications method that in turn calls an async GetPatientInfoAsync method:

1. Start by creating a new .NET console application

2. Add Patient, Provider, and Medication classes to a Models folder and HealthcareService and MedicationLoader classes to a SyncToAsync folder:

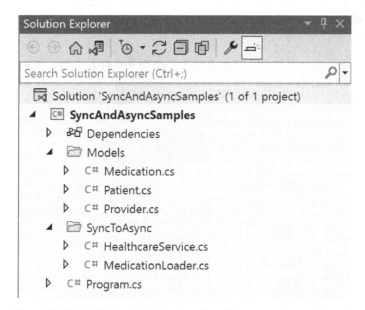

Figure 5.4: The initial project structure for calling async from sync code

3. Add the necessary properties for the model classes:

```
public class Medication
{
    public int Id { get; set; }
    public string? Name { get; set; }
}
public class Provider
{
    public int Id { get; set; }
    public string? Name { get; set; }
}
```

```
public class Patient
{
    public int Id { get; set; }
    public string? Name { get; set; }
    public List<Medication>? Medications { get; set; }
    public Provider? PrimaryCareProvider { get; set; }
}
```

4. Create the GetPatientInfoAsync method in the HealthcareService class. This method creates a patient with a provider and two medications after injecting a 2-second async delay:

```
public async Task<Patient> GetPatientInfoAsync
    (int patientId)
{
    await Task.Delay(2000);
    Patient patient = new()
    {
        Id = patientId,
        Name = "Smith, Terry",
        PrimaryCareProvider = new Provider
        {
            Id = 999,
            Name = "Dr. Amy Ng"
        },
        Medications = new List<Medication>
        {
            new Medication { Id = 1, Name =
                "acetaminophen" },
            new Medication { Id = 2, Name =
                "hydrocortisone cream" }
        }
    };
    return patient;
}
```

5. Add the implementation for the `MedicationLoader` service:

```
public class MedicationLoader
{
    private HealthcareService _healthcareService;
    public MedicationLoader()
    {
        _healthcareService = new HealthcareService();
    }
    public Patient? GetPatientAndMedications(int
        patientId)
    {
        Patient? patient = null;
        try
        {
            patient = _healthcareService
                .GetPatientInfoAsync(patientId).Result;
        }
        catch (AggregateException ae)
        {
            Console.WriteLine($"Error loading patient.
                Message: {ae.Flatten().Message}");
        }
        if (patient != null)
        {
            patient = ProcessPatientInfo(patient);
            return patient;
        }
        else
        {
            return null;
        }
    }
    private Patient ProcessPatientInfo(Patient
        patient)
```

```
        {
            // Add additional processing here.
            return patient;
        }
    }
```

The GetPatientAndMedications method calls GetPatientInfoAsync and uses the Result property to synchronously wait for the async method to complete and return the value. Using Result is the same as using the Wait() method on an async method that returns no value. The current thread is blocked until the method completes.

We have wrapped the call in a try/catch block that handles an AggregateException instance. If the call was successful, and the patient variable is not null, ProcessPatientInfo is called before returning the patient data to the caller.

6. Add this code to Program.cs to call the synchronous method:

```
using SyncAndAsyncSamples.Models;
using SyncAndAsyncSamples.SyncToAsync;
Console.WriteLine("Hello, sync to async world!");
var medLoader = new MedicationLoader();
Patient? patient = medLoader.GetPatientAndMedications
    (123);
Console.WriteLine($"Loaded patient: {patient.Name}
    with {patient.Medications.Count} medications.");
```

7. Run the program. You should see this output in the window:

```
Hello, sync to async world!
Loaded patient: Smith, Terry with 2 medications.
```

Next, let's try to load the same data but with an async method calling some legacy synchronous code.

Executing synchronous code as async

In this example, we will mirror the previous example. There will be a PatientLoader instance with async methods calling a PatientService instance with a synchronous method:

1. Add a PatientService class to a new AsyncToSync folder in your project.

2. Create a `GetPatientInfo` method with a similar implementation to the `GetPatientInfoAsync` method in the previous example:

```
public Patient GetPatientInfo(int patientId)
{
    Thread.Sleep(2000);
    Patient patient = new()
    {
        Id = patientId,
        Name = "Smith, Terry",
        PrimaryCareProvider = new Provider
        {
            Id = 999,
            Name = "Dr. Amy Ng"
        },
        Medications = new List<Medication>
        {
            new Medication { Id = 1, Name =
                "acetaminophen" },
            new Medication { Id = 2, Name =
                "hydrocortisone cream" }
        }
    };
    return patient;
}
```

The differences here are that the method is not `async`, it returns a `Patient` instance instead of a `Task<Patient>` instance, and we're injecting a delay with `Thread.Sleep` instead of `Task.Delay`.

3. Create the `PatientLoader` class in the `AsyncToSync` folder, and start its implementation by creating a new instance of `PatientService`:

```
private PatientService _patientService = new
    PatientService();
```

4. Now create the async version of `ProcessPatientInfo` from the previous example:

```
private async Task<Patient> ProcessPatientInfoAsync
    (Patient patient)
```

```
    {
        await Task.Delay(100);
        // Add additional processing here.
        return patient;
    }
```

5. Now create the GetPatientAndMedsAsync method:

```
public async Task<Patient?> GetPatientAndMedsAsync
    (int patientId)
{
    Patient? patient = null;
    try
    {
        patient = await Task.Run(() =>
            _patientService.GetPatientInfo(patientId));
    }
    catch (Exception e)
    {
        Console.WriteLine($"Error loading patient.
            Message: {e.Message}");
    }
    if (patient != null)
    {
        patient = await ProcessPatientInfoAsync
            (patient);
        return patient;
    }
    else
    {
        return null;
    }
}
```

The primary differences from the last example are highlighted. The synchronous class to GetPatientInfo is wrapped in a call to await Task.Run, which will wait for the call without blocking the current thread from performing other work.

We are now using `Exception` instead of `AggregateException` in the `catch` block. You should always use `AggregateException` with blocking `Wait` and `Result` calls and use `Exception` with `async` and `await`.

Finally, the async call to `ProcessPatientInfoAsync` is awaited if the `patient` variable is not `null`.

6. Next update `Program.cs` to call the new `PatientLoader` code:

```
using SyncAndAsyncSamples.AsyncToSync;
using SyncAndAsyncSamples.Models;
Console.WriteLine("Hello, async to sync world!");
var loader = new PatientLoader();
Patient? patient = await loader.GetPatientAndMedsAsync
    (123);
Console.WriteLine($"Loaded patient: {patient.Name}
    with {patient.Medications.Count} medications.");
```

7. Run the program, and the output should look similar to the previous example:

```
Hello, async to sync world!
Loaded patient: Smith, Terry with 2 medications.
```

By now, you should have a solid understanding of how to interop between asynchronous and synchronous code. Let's move forward and create an example of loading data from several `async` methods in parallel.

Working with multiple background tasks

In this section, we will see code samples for loading data from multiple sources in parallel, not waiting until the method is ready to return the data to the caller. The technique is slightly different for synchronous and asynchronous code, but the general idea is the same.

First, review this method that calls three async methods and uses `Task.WhenAll` to wait before returning the patient data:

```
public async Task<Patient> LoadPatientAsync(int patientId)
{
    var taskList = new List<Task>
    {
        LoadPatientInfoAsync(patientId),
        LoadProviderAsync(patientId),
```

```
                LoadMedicationsAsync(patientId)
        };
        await Task.WhenAll(taskList.ToArray());
        _patient.Medications = _medications;
        _patient.PrimaryCareProvider = _provider;
        return _patient;
    }
```

Now, review this synchronous version of the method, which uses `Task.WaitAll`:

```
    public Patient LoadPatient(int patientId)
    {
        var taskList = new List<Task>
        {
            LoadPatientInfoAsync(patientId),
            LoadProviderAsync(patientId),
            LoadMedicationsAsync(patientId)
        };
        Task.WaitAll(taskList.ToArray());
        _patient.Medications = _medications;
        _patient.PrimaryCareProvider = _provider;
        return _patient;
    }
```

Even this version of the code, which uses a blocking `WaitAll` call, will perform faster than making separate synchronous calls to the three methods.

The complete implementation of this `ParallelPatientLoader` class is available in the GitHub repository for this chapter. Let's finish up the chapter by listing some best practices for using `async`, `await`, and `Task` objects.

Asynchronous programming best practices

When working with async code, there are many best practices of which you should be aware. In this section, we will list the most important ones to remember in your day-to-day development. **David Fowler**, who is a veteran member of the ASP.NET team at Microsoft and a .NET expert, maintains an open source list of many other best practices. I recommend bookmarking this page for later reference while working with your own projects: `https://github.com/ davidfowl/AspNetCoreDiagnosticScenarios/blob/master/AsyncGuidance. md#asynchronous-programming`.

These are my top recommendations (in no particular order) to follow when working with async code:

1. Always prefer `async` and `await` over synchronous methods and blocking calls such as `Wait()` and `Result`. If you are creating a new project, you should build with async in mind from the start.

2. Unless you are using `Task.WhenAll` to wait for multiple operations simultaneously, you should directly await a method rather than creating a `Task` instance and awaiting it.

3. Do not use `async void`. Your async methods should always return Task, Task<TResult>, `ValueTask`, or `ValueTask<TResult>`. The only exceptions are event handlers that have existing signatures that return `void`. Event `Main` methods can be async in .NET 6.

4. Do not mix blocking code and asynchronous code. Use `async` calls through the call stack.

5. Use `Task.Run` instead of `Task.Factory.StartNew` unless you need to pass additional parameters to one of the `StartNew` overloaded methods.

6. Long-running `async` methods should support cancellation. We will discuss cancellation in depth in *Chapter 11*.

7. Synchronize the usage of shared data. Your code should add locks to prevent any overwriting of data in objects used across threads.

8. Always use `async` and `await` for I/O-bound work such as network and file access.

9. When you create an `async` method, add the Async suffix to its name. This helps to differentiate `sync` and `async` methods at a glance. An `async` method to return user information should be named `GetUserInfoAsync`, not `GetUserInfo`.

10. Do not use `Thread.Sleep` in async methods. If your code must wait for a fixed period, use `await Task.Delay`.

Those are my 10 rules to get you started, but there are many more best practices for async development with .NET. We will discover more of them as we progress through the remaining chapters.

Let's wrap up and review what we have learned about async programming in this chapter.

Summary

In this chapter, we have covered quite a bit of information about asynchronous development with C# and. NET. We started by covering some of the ways to handle I/O-bound and CPU-bound operations in your applications.

Next, we created some practical examples that use the `Task` and `Task<TResult>` classes and discovered how to work with multiple `Task` objects. You got some practical advice for interop between modern asynchronous code and legacy synchronous methods. Finally, we covered some of the most important rules to remember when working with asynchronous code and `Task` objects.

In the next chapter, *Chapter 6*, you will learn the ins and outs of parallel programming in .NET using the **Task Parallel Library** (**TPL**) and learn how to the avoid common pitfalls of parallel programming.

Questions

1. Which property of `Task` makes a blocking call to return data from the underlying method?

2. Which `async` method of the `Task` class should be used to await multiple tasks?

3. What is the blocking equivalent of `Task.WhenAll()`?

4. What type should an `async` method always return?

5. Are `async` methods more suited to I/O-bound or CPU-bound operations?

6. *True or false*: `Async` methods should never end with `Async` as their suffix.

7. What method can be used to wrap a synchronous method in an `async` call?

6

Parallel Programming Concepts

The **Task Parallel Library** (TPL) encompasses various .NET programming constructs, including parallel loops, parallel invocations, **PLINQ**, and task-based async programming. In *Chapter 5*, we explored async programming with Task objects. This chapter will delve deeper into the System. Threading.Tasks.Parallel members in the TPL and some additional tasking concepts for handling related tasks.

The lines between parallel programming, concurrency, and asynchronous programming are not always clear-cut, and you will discover where the three concepts intersect as we read ahead.

In this chapter, you will learn the following:

- Getting started with the TPL
- Parallel loops in .NET
- Relationships between parallel tasks
- Common pitfalls with parallelism

By the end of this chapter, you will understand how to use parallel programming in your own projects, why you would choose a parallel loop over a standard loop, and when to use async and await instead of a parallel loop.

Technical requirements

To follow along with the examples in this chapter, the following software is recommended for Windows developers:

- Visual Studio 2022 version 17.0 or later
- .NET 6

While these are recommended, if you have .NET 6 installed, you can use your preferred editor. For example, Visual Studio 2022 for Mac on macOS 10.13 or later, JetBrains Rider, or Visual Studio Code will work just as well.

All the code examples for this chapter can be found on GitHub at `https://github.com/ PacktPublishing/Parallel-Programming-and-Concurrency-with-C- sharp-10-and-.NET-6/tree/main/chapter06`.

Let's get started by discussing the TPL and where it fits within the world of parallel programming in .NET.

Getting started with the TPL

The **TPL** consists of the types that were added to the `System.Threading` and `System. Threading.Tasks` namespaces in *.NET Framework 4.0*. The TPL provides features that make parallelism and concurrency simpler for .NET developers. There is no need to manage the `ThreadPool` tasks in your code. The TPL handles thread management and automatically scales the number of active threads based on processor capability and availability.

Developers should use the TPL when they need to introduce parallelism or concurrency to their code for improved performance. However, the TPL is not the right choice for every scenario. How do you know when to choose the TPL and which TPL constructs are the best choice for each scenario?

Let's explore a few common scenarios.

I/O-bound operations

When dealing with I/O-bound operations such as file operations, database calls, or web service calls, asynchronous programming with `Task` objects and C# `async`/`await` operations are your best choice. If your service requires that you loop through a large collection, making a service call for each object in the loop, you should consider refactoring the service to return the data as a single service call. This will minimize the overhead associated with each network operation. It will also allow your client code to make a single `async` call to the service while keeping the main thread free to do other work.

I/O-bound operations are usually not suited to parallel operations, but there are exceptions to every rule. If you need to iterate through a set of folders and subfolders in the filesystem, a parallel loop can be well-suited for this. However, it is important that none of the iterations of your loop attempt to access the same file in order to avoid locking issues.

Now, let's explore some CPU-bound scenarios.

CPU-bound operations

CPU-bound operations are not reliant on outside resources such as the filesystem, network, or the internet. They involve processing data in memory within your application's process. There are many types of data transformation that fall into this category. Your application may be serializing or deserializing data, converting between file types, or processing images or other binary data.

These types of operations make sense for data parallelism and parallel loops in particular, with a couple of exceptions. First, if each iteration is not very CPU intensive, using the TPL is not worth the overhead it introduces. If the process is very intensive, but there are very few objects to iterate over, consider using `Parallel.Invoke` instead of one of the parallel loops, `Parallel.For` or `Parallel.ForEach`. Using parallel constructs for less CPU-intense operations can often slow your code due to the overhead of using the TPL. In *Chapter 10* we will learn how to use Visual Studio to determine the performance of parallel and concurrent code.

Now that you have some understanding of when to use parallelism in your applications, let's explore some practical examples of using `Parallel.For` and `Parallel.ForEach`.

Parallel loops in .NET

In this section, we will explore some examples of leveraging data parallelism in .NET projects. The parallel versions of the C# `for` and `foreach` loops, `Parallel.For` and `Parallel.ForEach`, are part of the `System.Threading.Tasks.Parallel` namespace. Using these parallel loops is similar to using their standard counterparts in C#.

One key difference is that the body of the parallel loops is declared as a **lambda expression**. As a result, there are some changes to how you would continue or break from the parallel loops. Instead of using `continue` to stop the current iteration of the loop without breaking the entire loop, you would use a `return` statement. The equivalent of using `break` to break out of a parallel loop is to use the `Stop()` or `Break()` statements.

Let's look at an example of using a `Parallel.For` loop in a .NET WinForms application.

Basic Parallel.For loops

We are going to create a new **WinForms application** that allows users to select a folder on their workstation and examine some information about the files in the selected folder. The project's `FileProcessor` class will iterate the files to aggregate the file size and find the most recently written file:

1.   Start by creating a new .NET 6 WinForms project in Visual Studio

2. Add a new class named `FileData`. This class will contain the data from `FileProcessor`:

```
public class FileData
{
    public List<FileInfo> FileInfoList { get; set; } =
        new();
    public long TotalSize { get; set; } = 0;
    public string LastWrittenFileName
        { get; set; } = "";
    public DateTime LastFileWriteTime { get; set; }
}
```

We will be returning a list of the `FileInfo` objects for the files in the selected folder, the total size of all files, the name of the last written file, and the date and time that the file was written.

3. Next, create a new class named `FileProcessor`

4. Add a static method named `GetInfoForFiles` to `FileProcessor`:

```
public static FileData GetInfoForFiles(string[] files)
{
    var results = new FileData();
    var fileInfos = new List<FileInfo>();
    long totalFileSize = 0;
    DateTime lastWriteTime = DateTime.MinValue;
    string lastFileWritten = "";
    object dateLock = new();
    Parallel.For(0, files.Length,
            index => {
                FileInfo fi = new(files[index]);
                long size = fi.Length;
                DateTime lastWrite =
                    fi.LastWriteTimeUtc;
                lock (dateLock)
                {
                    if (lastWriteTime < lastWrite)
                    {
                        lastWriteTime = lastWrite;
                        lastFileWritten = fi.Name;
                    }
```

```
            }
            Interlocked.Add(ref totalFileSize,
                size);
            fileInfos.Add(fi);
        });
    results.FileInfoList = fileInfos;
    results.TotalSize = totalFileSize;
    results.LastFileWriteTime = lastWriteTime;
    results.LastWrittenFileName = lastFileWritten;
    return results;
}
```

The `Parallel.For` loop and the **lambda expression** of its body are highlighted in the preceding code. There are a few things to note about the code inside the loop:

I.   First, `index` is provided as a parameter to the lambda expression so the expression body can use it to access the current member of the `files` array.

II.   The `totalFileSize` gets updated inside a call to `Interlocked.Add`. This is the most efficient way to safely add values in parallel code.

III. There isn't a simple way to leverage `Interlocked` to update the `lastWriteTime` DateTime value. So, instead, we are using a `lock` block with a `dateLock` object to safely read and set the `lastWriteTime` method-level variable.

5. Next, open the designer for `Form1.cs` and add the following controls to the form:

```
private GroupBox FileProcessorGroup;
private Button FolderProcessButton;
private Button FolderBrowseButton;
private TextBox FolderToProcessTextBox;
private Label label1;
private TextBox FolderResultsTextBox;
private Label label2;
private FolderBrowserDialog folderToProcessDialog;
```

View the `Form1.designer.cs` file on this chapter's GitHub repository (https://github.com/PacktPublishing/Parallel-Programming-and-Concurrency-with-C-sharp-10-and-.NET-6/tree/main/chapter06/WinFormsParallelLoopApp) to review and set all of the properties for these controls.

When you are finished, the form's designer should look like this:

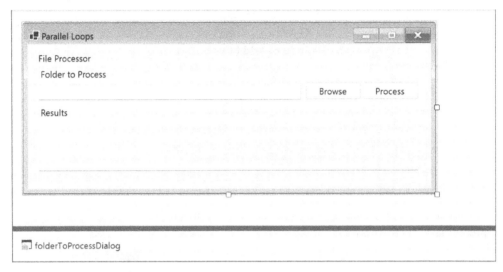

Figure 6.1 – The completed Form1.cs designer view in Visual Studio

6. Next, double-click the **Browse** button in the `Form1` designer, and a `FolderBrowserButton_` `Click` event handler will be generated in the code-behind file. Add the following code to use the `folderToProcessDialog` object to show a folder picker dialog to the user:

```
private void FolderBrowseButton_Click(object sender,
    EventArgs e)
{
    var result = folderToProcessDialog.ShowDialog();
    if (result == DialogResult.OK)
    {
        FolderToProcessTextBox.Text =
            folderToProcessDialog.SelectedPath;
    }
}
```

The selected folder path will be set in `FolderToProcessTextBox` for use in the next step. The user can alternatively manually type or paste a folder path in the field. If you would like to prevent manual entry, you can set `FolderToProcessTextBox.ReadOnly` to `true`.

7. Next, double-click the **Process** button in the designer view. A `FolderProcessButton_Click` event handler will be generated in the code behind. Add the following code to call `FileProcessor` and display the results in `FolderResultsTextBox`:

```
private void FolderProcessButton_Click(object sender,
    EventArgs e)
{
    if (!string.IsNullOrWhiteSpace
        (FolderToProcessTextBox.Text) &&
        Directory.Exists(FolderToProcessTextBox.Text))
    {
        string[] filesToProcess = Directory.GetFiles
            (FolderToProcessTextBox.Text);
        FileData? results = FileProcessor
            .GetInfoForFiles(filesToProcess);
        if (results == null)
        {
            FolderResultsTextBox.Text = "";
            return;
        }
        StringBuilder resultText = new();
        resultText.Append($"Total file count:
            {results.FileInfoList.Count}; ");
        resultText.AppendLine($"Total file size:
            {results.TotalSize} bytes");
        resultText.Append($"Last written file:
            {results.LastWrittenFileName} ");
        resultText.Append($"at
            {results.LastFileWriteTime}");
        FolderResultsTextBox.Text =
            resultText.ToString();
    }
}
```

The code here is straightforward enough. The static `GetInfoForFiles` method returns a `FileData` instance with the file information. We're using `StringBuilder` to create the output to be set in `FolderResultsTextBox`.

8. We're ready to run the application. Start debugging the project in Visual Studio and give it a try. Your results should look something like this:

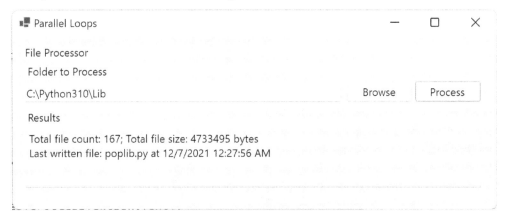

Figure 6.2 – Running the Parallel Loops application

That's all there is to it. If you want to try something more advanced, you can try modifying the project to also process files in all subfolders of the selected folder. Let's make a different change to the project so we can reduce the locking calls to `Interlocked.Add`.

Parallel loops with thread-local variables

The `Parallel.For` construct has an overload that will allow our code to keep a running subtotal of the total file size for each thread participating in the loop. What that means is that we will only need to use `Interlocked.Add` when aggregating the subtotal from each thread to `totalFileSize`. This is accomplished by providing a **thread-local variable** to the loop. The subtotal in the following code is stored discretely for each thread. So, if the loop has 200 iterations, but only 5 threads participate in the loop, `Interlocked.Add` will only be called 5 times instead of 200 times without losing any thread safety:

```
public static FileData GetInfoForFilesThreadLocal(string[]
    files)
{
    var results = new FileData();
    var fileInfos = new List<FileInfo>();
    long totalFileSize = 0;
    DateTime lastWriteTime = DateTime.MinValue;
    string lastFileWritten = "";
```

```
    object dateLock = new();
    Parallel.For<long>(0, files.Length, () => 0,
        (index, loop, subtotal) => {
            FileInfo fi = new(files[index]);
            long size = fi.Length;
            DateTime lastWrite = fi.LastWriteTimeUtc;
            lock (dateLock)
            {
                if (lastWriteTime < lastWrite)
                {
                    lastWriteTime = lastWrite;
                    lastFileWritten = fi.Name;
                }
            }
            subtotal += size;
            fileInfos.Add(fi);
            return subtotal;
            },
        (runningTotal) => Interlocked.Add(ref
            totalFileSize, runningTotal)
    );
    results.FileInfoList = fileInfos;
    results.TotalSize = totalFileSize;
    results.LastFileWriteTime = lastWriteTime;
    results.LastWrittenFileName = lastFileWritten;
    return results;
}
```

To summarize the preceding changes, you will notice we are using the Parallel.For<long> generic method to indicate that the subtotal thread-local variable should be long instead of int (the default type). The size is added to subtotal in the first lambda expression without any locking expression. We now have to return subtotal, so the other iterations have access to the data. Finally, we have added a final parameter to For with a lambda expression that adds each thread's runningTotal to totalFileSize using Interlocked.Add.

If you update FolderProcessButton_Click to call GetInfoForFilesThreadLocal, the output will be the same, but the performance will be improved, perhaps not noticeably. The performance improvement depends on the number of files in your selected folder.

Now that we have tried a couple of exercises with the `Parallel.For` loop, let's create a sample using the `Parallel.ForEach` method.

Simple Parallel.ForEach loops

The `Parallel.ForEach` methods, such as `Parallel.For`, are similar in use to their non-parallel counterpart. You would use `Parallel.ForEach` over `Parallel.For` when you have an `IEnumerable` collection to process. In this sample, we will create a new method that accepts a `List<string>` of image files to iterate and convert to `Bitmap` objects:

1. Start by creating a new private static method named `ConvertJpgToBitmap` in the `FileProcessor` class. This method will open each JPG file and return a new `Bitmap` containing the image data:

    ```
    private static Bitmap ConvertJpgToBitmap(string
        fileName)
    {
        Bitmap bmp;
        using (Stream bmpStream = File.Open(fileName,
            FileMode.Open))
        {
            Image image = Image.FromStream(bmpStream);
            bmp = new Bitmap(image);
        }
        return bmp;
    }
    ```

2. Next, create a public static method in the same class named `ConvertFilesToBitmaps`:

    ```
    public static List<Bitmap> ConvertFilesToBitmaps
        (List<string> files)
    {
        var result = new List<Bitmap>();
        Parallel.ForEach(files, file =>
        {
            FileInfo fi = new(file);
            string ext = fi.Extension.ToLower();
            if (ext == ".jpg" || ext == ".jpeg")
    ```

```
        {
            result.Add(ConvertJpgToBitmap(file));
        }
    });
    return result;
}
```

This method accepts List<string> containing the files in the selected folder. Inside the Parallel.ForEach loop, it checks whether the file has a .jpg or .jpeg file extension. If it does, it is converted to a bitmap and added to the result collection.

3. Add a new button to Form1.cs. Set the Name property as ProcessJpgsButton and the Text property as Process JPGs.

4. Double-click the new button to create an event handler in the code-behind file. Add the following code to the new event handler:

```
private void ProcessJpgsButton_Click(object sender,
    EventArgs e)
{
    if (!string.IsNullOrWhiteSpace
        (FolderToProcessTextBox.Text) &&
        Directory.Exists(FolderToProcessTextBox.Text))
    {
        List<string> filesToProcess = Directory
            .GetFiles(FolderToProcessTextBox.Text)
                .ToList();
        List<Bitmap> results = FileProcessor
            .ConvertFilesToBitmaps(filesToProcess);
        StringBuilder resultText = new();
        foreach (var bmp in results)
        {
            resultText.AppendLine($"Bitmap height:
                {bmp.Height}");
        }
        FolderResultsTextBox.Text =
            resultText.ToString();
    }
}
```

5. Now, run the project, select a folder containing some JPG files, and click the new **Process JPGs** button. You should see the height of each converted JPG listed in the output.

That's all you need for a simple `Parallel.ForEach` loop. What can you do if you need to cancel a long-running parallel loop? Let's update our example to do just that with `Parallel.ForEachAsync`.

Cancel a Parallel.ForEachAsync loop

`Parallel.ForEachAsync` is new in .NET 6. It is an awaitable version of `Parallel.ForEach` with an `async` lambda expression as its body. Let's update the previous example to use this new parallel method and add the ability to cancel the operation:

1. We are going to start by making an `async` copy of `ConvertFilesToBitmaps` named `ConvertFilesToBitmapsAsync`. The differences are highlighted in the following:

```
public static async Task<List<Bitmap>>
    ConvertFilesToBitmapsAsync(List<string> files,
        CancellationTokenSource cts)
{
    ParallelOptions po = new()
    {
        CancellationToken = cts.Token,
        MaxDegreeOfParallelism =
            Environment.ProcessorCount == 1 ? 1
                        : Environment.ProcessorCount - 1
    };
    var result = new List<Bitmap>();
    try
    {
        await Parallel.ForEachAsync(files, po, async
            (file, _cts) =>
        {
            FileInfo fi = new(file);
            string ext = fi.Extension.ToLower();
            if (ext == ".jpg" || ext == "jpeg")
            {
                result.Add(ConvertJpgToBitmap(file));
                await Task.Delay(2000, _cts);
```

```
        }
      });
  }
  catch (OperationCanceledException e)
  {
      MessageBox.Show(e.Message);
  }
  finally
  {
      cts.Dispose();
  }
  return result;
}
```

The new method is async, returns Task<List<Bitmap>>, accepts
CancellationTokenSource, and uses that when creating ParallelOptions
to pass to the Parallel.ForEachAsync method. Parallel.ForEachAsync
is awaited and its lambda expression is declared as async so we can await the new Task.
Delay that has been added to give us enough time to click the **Cancel** button before the
loop completes.

Enclosing Parallel.ForEachAsync in a try/catch block that handles
OperationCanceledException enables the method to catch the cancellation. We'll
show a message to the user after the cancellation is handled.

The code is also setting the ProcessorCount option. If there is only one CPU core
available, we will set the value to 1; otherwise, we want to use no more than the number of
available cores minus one. The .NET runtime typically manages this value very well, so you
should only change this option if you find it improves your application's performance.

2. In the Form1.cs file, add a new CancellationTokenSource private variable:

    ```
    private CancellationTokenSource _cts;
    ```

3. Update the event handler to be async, set _cts to be a new instance of
    CancellationTokenSource, and pass it to ConvertFilesToBitmapsAsync.
    Add await to that call as well.

    All of the necessary changes are highlighted in the following snippet:

    ```
    private async void ProcessJpgsButton_Click(object
        sender, EventArgs e)
    {
        if (!string.IsNullOrWhiteSpace
    ```

```
            (FolderToProcessTextBox.Text) &&
            Directory.Exists(FolderToProcessTextBox.Text))
        {

            _cts = new CancellationTokenSource();
            List<string> filesToProcess = Directory
                .GetFiles(FolderToProcessTextBox.Text)
                    .ToList();
            List<Bitmap> results = await FileProcessor
                .ConvertFilesToBitmapsAsync
                    (filesToProcess, _cts);
            StringBuilder resultText = new();
            foreach (var bmp in results)
            {
                resultText.AppendLine($"Bitmap height:
                    {bmp.Height}");
            }
            FolderResultsTextBox.Text = resultText
                .ToString();
        }
    }
```

4. Add a new button to the form named `CancelButton` with a caption of `Cancel`

5. Double-click the **Cancel** button and add the following event handler code:

```
    private void CancelButton_Click(object sender,
        EventArgs e)
    {
        if (_cts != null)
        {
            _cts.Cancel();
        }
    }
```

6. Run the application, browse to and select a folder containing JPG files, click the **Process JPGs** button, and immediately click the **Cancel** button. You should receive a message indicating that processing has been canceled. No further records are processed.

We will learn more about canceling asynchronous and parallel work in *Chapter 11*. Now, let's discuss the `Parallel.Invoke` construct and relationships between tasks in the TPL.

Relationships between parallel tasks

In the previous chapter, *Chapter 5*, we learned how to use `async` and `await` to perform work in parallel and manage the flow of tasks by using `ContinueWith`. In this section, we will examine some of the TPL features that can be leveraged to manage relationships between tasks running in parallel.

Let's start by looking deeper into the `Parallel.Invoke` method provided by the TPL.

Under the covers of Parallel.Invoke

In *Chapter 2*, we learned how to use the `Parallel.Invoke` method to execute multiple tasks in parallel. We are going to revisit `Parallel.Invoke` now and discover what is happening under the covers. Consider using it to invoke two methods:

```
Parallel.Invoke(DoFirstAction, DoSectionAction);
```

This is what is happening behind the scenes:

```
List<Task> taskList = new();
taskList.Add(Task.Run(DoFirstAction));
taskList.Add(Task.Run(DoSectionAction));
Task.WaitAll(taskList.ToArray());
```

Two tasks will be created and queued on the thread pool. Assuming the system has available resources, the two tasks should be picked up and run in parallel. The calling method will block the current thread, waiting for the parallel tasks to complete. The action will block the calling thread for the duration of the longest-running task.

If this is acceptable for your application, using `Parallel.Invoke` makes the code cleaner and easy to understand. However, if you don't want to block the calling thread, there are a couple of options. First, let's make a change to the second example to use `await`:

```
List<Task> taskList = new();
taskList.Add(Task.Run(DoFirstAction));
taskList.Add(Task.Run(DoSectionAction));
await Task.WhenAll(taskList.ToArray());
```

By awaiting `Task.WhenAll` instead of using `Task.WaitAll`, we're allowing the current thread to do other work while waiting for the two child tasks to finish processing in parallel. To achieve the same result with `Parallel.Invoke`, we can wrap it in `Task`:

```
await Task.Run(() => Parallel.Invoke(DoFirstTask,
    DoSecondTask));
```

The same technique can be used with `Parallel.For` to avoid blocking the calling thread while waiting for the loop to complete. This is not necessary for `Parallel.ForEach`. Instead of wrapping `Parallel.ForEach` in `Task`, we can replace it with `Parallel.ForEachAsync`. We learned earlier in this chapter that .NET 6 added `Parallel.ForEachAsync`, which returns `Task` and can be awaited.

Next, let's discuss how the relationship between parent tasks and their children can be managed.

Understanding parallel child tasks

When executing nested tasks, by default, the parent task will not wait for its child tasks unless we use the `Wait()` method or `await` statements. However, this default behavior can be controlled with some options when using `Task.Factory.StartNew()`. To illustrate the available options, we are going to create a new sample project:

1. First, create a new C# console application named `ParallelTaskRelationshipsSample`.

2. Add a class to the project named `ParallelWork`. This is where we will create the parent methods and their children.

3. Add the three following methods to the `ParallelWork` class. These will be our child methods. Each one writes some console output when starting and completing. Delays are injected with `Thread.SpinWait`. If you are unfamiliar with `Thread.SpinWait`, it puts the current thread into a loop for the number of iterations specified, injecting a wait without removing the thread from consideration with the scheduler:

    ```
    public void DoFirstItem()
    {
        Console.WriteLine("Starting DoFirstItem");
        Thread.SpinWait(1000000);
        Console.WriteLine("Finishing DoFirstItem");
    }
    public void DoSecondItem()
    {
    ```

```
        Console.WriteLine("Starting DoSecondItem");
        Thread.SpinWait(1000000);
        Console.WriteLine("Finishing DoSecondItem");
    }
    public void DoThirdItem()
    {
        Console.WriteLine("Starting DoThirdItem");
        Thread.SpinWait(1000000);
        Console.WriteLine("Finishing DoThirdItem");
    }
```

4. Next, add a method named `DoAllWork`. This method will create a parent task that calls the preceding three methods with child tasks. There is no code added to wait for the child tasks:

```
    public void DoAllWork()
    {
        Console.WriteLine("Starting DoAllWork");
        Task parentTask = Task.Factory.StartNew(() =>
        {
            var child1 = Task.Factory.StartNew
                (DoFirstItem);
            var child2 = Task.Factory.StartNew
                (DoSecondItem);
            var child3 = Task.Factory.StartNew
                (DoThirdItem);
        });
        parentTask.Wait();
        Console.WriteLine("Finishing DoAllWork");
    }
```

5. Now, add some code to run `DoAllWork` from `Program.cs`:

```
    using ParallelTaskRelationshipsSample;
    var parallelWork = new ParallelWork();
    parallelWork.DoAllWork();
    Console.ReadKey();
```

6.   Run the program and examine the output. As you might expect, the parent task completes before its children:

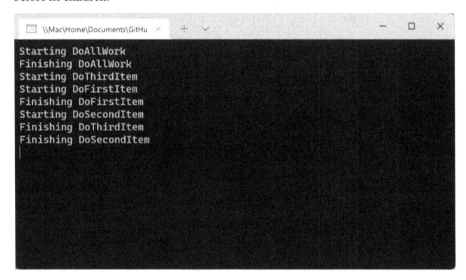

Figure 6.3 – The console application runs DoAllWork

7.   Next, let's create a method named `DoAllWorkAttached`. This method will run the same three child tasks, but the child task will include the `TaskCreationOptions.AttachedToParent` option:

```
public void DoAllWorkAttached()
{
    Console.WriteLine("Starting DoAllWorkAttached");
    Task parentTask = Task.Factory.StartNew(() =>
    {
        var child1 = Task.Factory.StartNew
            (DoFirstItem, TaskCreationOptions
                .AttachedToParent);
        var child2 = Task.Factory.StartNew
            (DoSecondItem, TaskCreationOptions
                .AttachedToParent);
        var child3 = Task.Factory.StartNew
            (DoThirdItem, TaskCreationOptions
                .AttachedToParent);
    });
    parentTask.Wait();
```

```
            Console.WriteLine("Finishing DoAllWorkAttached");
    }
```

8. Update `Program.cs` to call `DoAllWorkAttached` instead of `DoAllWork` and run the application again:

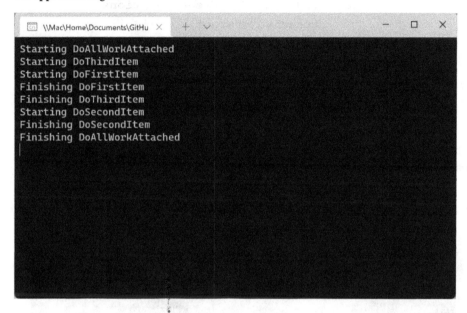

Figure 6.4 – Running our application and calling DoAllWorkAttached

You can see that even though we are not explicitly waiting for the child tasks, the parent task does not complete until its children do.

Now, suppose you have another parent that should not wait for its child tasks, regardless of whether they are started with the `TaskCreationOptions.AttachedToParent` option. Let's create a new method that can handle this scenario:

1. Create a method named `DoAllWorkDenyAttach` with the following code:

```
public void DoAllWorkDenyAttach()
{
    Console.WriteLine("Starting DoAllWorkDenyAttach");
    Task parentTask = Task.Factory.StartNew(() =>
    {
        var child1 = Task.Factory.StartNew
            (DoFirstItem, TaskCreationOptions
                .AttachedToParent);
```

```
                var child2 = Task.Factory.StartNew
                    (DoSecondItem, TaskCreationOptions
                        .AttachedToParent);
                var child3 = Task.Factory.StartNew
                    (DoThirdItem, TaskCreationOptions
                        .AttachedToParent);
            }, TaskCreationOptions.DenyChildAttach);
            parentTask.Wait();
            Console.WriteLine("Finishing DoAllWork
                DenyAttach");
        }
```

The child tasks are still being created with the AttachedToParent option, but the parent task now has a DenyChildAttach option set. This will supersede the child requests to attach to the parent.

2. Update Program.cs to call DoAllWorkDenyAttach and run the application once more:

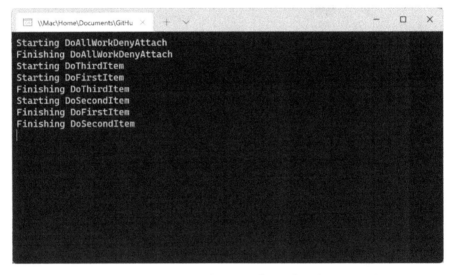

Figure 6.5 – The console application calls DoAllWorkDenyAttach

You can see that DenyChildAttach did override the AttachToParent option set on each child task. The parent completed without waiting for the children, as it did when calling DoAllWork.

One final note about this example. You may have noticed that we used `Task.Factory.StartNew` instead of `Task.Run`, even when we didn't need to set `TaskCreationOption`. That is because `Task.Run` will prohibit any child tasks from attaching to a parent. If you used `Task.Run` for the parent task in the `DoAllWorkAttached` method, the parent would have completed first, as it did in the other methods.

Let's finish up this chapter by covering some potential pitfalls when working with parallel programming in .NET.

Common pitfalls with parallelism

When working with the TPL, there are some practices to avoid in order to ensure the best outcomes in your applications. In some cases, parallelism used incorrectly can result in performance degradation. In other cases, it can cause errors or data corruption.

Parallelism is not guaranteed

When using one of the parallel loops or `Parallel.Invoke`, the iterations can run in parallel, but they are not guaranteed to do so. The code in these parallel delegates should be able to run successfully in either scenario.

Parallel loops are not always faster

We discussed this earlier in this chapter, but it is important to remember that parallel versions of `for` and `foreach` loops are not always faster. If each loop iteration runs quickly, the overhead of adding parallelism can slow down your application.

This is important to remember when introducing any threading to applications. Always test your code before and after introducing concurrency or parallelism to ensure that the performance gains are worth the overhead of threading.

Beware of blocking the UI thread

Remember that `Parallel.For` and `Parallel.ForEach` are *blocking calls*. If you use them on the UI thread, they will block the UI for the duration of the call. This blocking duration will be, at a minimum, the duration of the longest-running loop iteration.

As we discussed in the previous section, you can wrap the parallel code in a call to `Task.Run` to move the execution from the UI thread to a background thread on the thread pool.

Thread safety

Do not make calls to .NET methods that are not thread-safe within parallel loops. The thread safety of each .NET type is documented on Microsoft Docs. Use the .NET API browser to quickly find information about specific .NET APIs: `https://docs.microsoft.com/dotnet/api/`.

Limit the use of static .NET methods in parallel loops, even if they are marked as thread-safe. They will not cause errors or problems with data consistency, but they can negatively impact the loop performance. Even calls to `Console.WriteLine` should only be used for testing or demonstration purposes. Do not use these in production code.

UI controls

In Windows client applications, do not try to access UI controls within parallel loops. WinForms and WPF controls can only be accessed from the thread on which they were created. You can use `Dispatcher.Invoke` to invoke actions on other threads, but this will have performance implications. It is best to update the UI after your parallel loops have been completed.

ThreadLocal data

Remember to take advantage of `ThreadLocal` variables in your parallel loops. We illustrated how to do this in the *Parallel loops with thread-local variables* section earlier in this chapter.

That covers your introduction to parallel programming with C# and .NET. Let's wrap up by reviewing everything we have learned in the chapter.

Summary

In this chapter, we learned how to leverage parallel programming concepts in our .NET applications. We got hands-on with `Parallel.For`, `Parallel.ForEach`, and `Parallel.ForEachAsync` loops. In those sections, we learned how to safely aggregate data while maintaining thread safety. Next, we learned how to manage relationships between parent tasks and their parallel children. This will help to ensure your applications maintain an expected order of operations.

Finally, we covered some important pitfalls to avoid when implementing parallelism in our applications. Developers will want to pay close attention to avoid any of these pitfalls in their own applications.

To read more about data parallelism in .NET, the *Data Parallelism* documentation on Microsoft Docs is a great place to start: `https://docs.microsoft.com/dotnet/standard/parallel-programming/data-parallelism-task-parallel-library`.

In the next chapter, we will continue our exploration of the TPL by learning how to leverage the various building blocks included in the TPL Dataflow Library.

Questions

1. Which parallel loop executes a delegate in parallel for a given number of iterations?

2. Which parallel loop is the awaitable version of `Parallel.ForEach`?

3. Which parallel method can execute two or more provided actions in parallel?

4. Which `Task.Factory.StartNew` option can attach a child task's completion to its parent?

5. Which `Task.Factory.StartNew` option can be provided to a parent task to prevent any child tasks from attaching?

6. Why should you never use `Task.Run` when using `TaskCreationOptions` to establish parent/child relationships?

7. Are parallel loops always faster than their traditional counterparts?

7

Task Parallel Library (TPL) and Dataflow

The **Task Parallel Library (TPL) dataflow library** contains building blocks to orchestrate asynchronous workflows in .NET. This chapter will introduce the TPL Dataflow library, describe the types of **dataflow blocks** in the library, and illustrate some common patterns for using dataflow blocks through hands-on examples.

The dataflow library can be useful when processing large amounts of data in multiple stages or when your application receives data in a continuous stream. The dataflow blocks provide a fantastic way of implementing the **producer/consumer design pattern**.

To understand this, we will create a sample project that implements this pattern and examine other real-world uses of the dataflow library.

> **Note**
> It's important to know that the TPL Dataflow library isn't distributed as part of the .NET runtime or SDK. It's available as a NuGet package from Microsoft. We will add it to our sample projects with **NuGet Package Explorer (NPE)** in Visual Studio.

In this chapter, we will cover the following topics:

- Introducing the TPL Dataflow library
- Implementing the producer/consumer pattern
- Creating a data pipeline with multiple blocks
- Manipulating data from multiple data sources

By the end of this chapter, you will understand the purpose of each type of dataflow block and be able to add the dataflow library to your projects, where appropriate.

You will also know when dataflow blocks do not provide an advantage over simpler parallel programming alternatives, such as `Parallel.ForEach`.

Technical requirements

To follow along with the examples in this chapter, the following software is recommended for Windows developers:

- Visual Studio 2022 version 17.0 or later

- .NET 6

- To complete the WPF sample, you will need to install the .NET desktop development workload for Visual Studio

While these are recommended, if you have .NET 6 installed, you can use your preferred editor. For example, Visual Studio 2022 for Mac on macOS 10.13 or later, JetBrains Rider, or Visual Studio Code will work just as well.

The code examples for this chapter can be found on GitHub at `https://github.com/ PacktPublishing/Parallel-Programming-and-Concurrency-with-C- sharp-10-and-.NET-6/tree/main/chapter07`.

Let's get started by discussing the TPL Dataflow library and why it can be a great way to implement parallel programming in .NET.

Introducing the TPL Dataflow library

The TPL Dataflow library has been available for as long as TPL itself. It was released in 2010 after **.NET Framework 4.0** reached its RTM milestone. The members of the dataflow library are part of the `System.Threading.Tasks.Dataflow` namespace. The dataflow library is intended to build on the basics of parallel programming that are provided in TPL, expanding to address data flow scenarios (hence the name of the library). The dataflow library is made up of foundational classes called **blocks**. Each data flow block is responsible for a particular action or step in the overall flow.

The dataflow library consists of three basic types of blocks:

- **Source blocks**: These blocks implement the `ISourceBlock<TOutput>` interface. Source blocks can have their data read from the workflow you define.

- **Target blocks**: This type of block implements the `ITargetBlock<TInput>` interface and is a data receiver.

- **Propagator blocks**: These blocks act as both source and target. They implement the `IPropagatorBlock<TInput, TOutput>` interface. Applications can read data from these blocks and write to them.

When you connect multiple dataflow blocks to create a workflow, the resulting system is referred to as a **dataflow pipeline**. You can connect a source block to a target block with the `ISourceBlock<TOutput>.LinkTo` method. This is where propagator blocks can fit in the middle of a pipeline. They can act as both the source and target of a link in the workflow. If a message from a source block can be processed by more than one target, you can add filtering to examine the properties of the object provided by the source to determine which target or propagator block should receive the object.

The objects that are passed between dataflow blocks are commonly referred to as **messages**. You can think of a dataflow pipeline as a **network** or messaging system. The units of data that flow through the network are the messages. Each block is responsible for reading, writing, or transforming each message in some way.

To send a message to a target block, you can use the `Post` method to send it synchronously or the `SendAsync` method to send it asynchronously. In source blocks, messages can be received with the `Receive`, `TryReceive`, and `ReceiveAsync` methods. The `Receive` and `TryReceive` methods are both synchronous. The `Choose` method will monitor multiple source blocks for data and return a message from the first source to provide data.

To offer a message from a source block to a target block, the source can call the `OfferData` method of a target. The `OfferData` method returns a `DataflowMessageStatus` enum that has several possible values:

- `Accepted`: The message was accepted and will be processed by the target.
- `Declined`: The message was declined by the target. The source block still owns the message and cannot process its next message until the current message has been accepted by another target.
- `DecliningPermanently`: The message was declined, and the target is no longer available for processing. All subsequent messages will be declined by the current target. Source blocks will unlink from a target that returns this status.
- `Postponed`: Accepting the message has been postponed. It may be accepted by the target at a later time. In this case, the source can wait or attempt to pass the message to an alternative target block.
- `NotAvailable`: The message was no longer available when the target tried to accept it. This can occur when the target attempts to accept a message after it had been postponed, but the source block has already passed the message to a different target block.

Dataflow blocks support the concept of **completion** by providing a `Complete` method and a `Completion` property. The `Complete` method is called to request completion on a block, while the `Completion` property returns a `Task`, known as the block's **completion task**. These completion members are part of the `IDataflowBlock` interface, which is inherited by both `ISourceBlock` and `ITargetBlock`.

The completion task can be used to determine if a block has encountered an error or has been canceled. Let's see how:

1. The simplest way to handle errors encountered by a dataflow block is to call `Wait` on the `Completion` property of the block and handle the `AggregateException` exception type in the `try`/`catch` block:

```
try
{
    inputBlock.Completion.Wait();
}
catch (AggregateException ae)
{
    ae.Handle(e =>
    {
        Console.WriteLine($"Error processing input -
            {e.GetType().Name}: {e.Message}");
    });
}
```

2. If you want to do the same thing without using the blocking `Wait` call, you can `await` the completion task and handle the `Exception` type:

```
try
{
    await inputBlock.Completion;
}
catch (Exception e)
{
    Console.WriteLine($"Error processing input -
        {e.GetType().Name}: {e.Message}");
}
```

3. Another alternative is to use the `ContinueWith` method on the completion task. Inside the continuation block, you can check the status of the task to determine if it is `Faulted` or `Canceled`:

```
try
{
    inputBlock.ContinueWith(task =>
```

```
        {
                Console.WriteLink($"Task completed with a
                    status of {task.Status}");
        });
        await inputBlock.Completion;
    }
    catch (Exception e)
    {
        Console.WriteLine($"Error processing input -
            {e.GetType().Name}: {e.Message}");
    }
```

We will see more comprehensive examples of dataflow block use when we create a sample project using the producer/consumer pattern in the next section. Before we examine the types of dataflow blocks, let's discuss why Microsoft created the library.

Why use the TPL Dataflow library?

The TPL dataflow library was created by Microsoft as a means of orchestrating asynchronous data processing workflows. Data flows into the first dataflow block in the pipeline from a data source. The source can be a database, a local or network folder, a camera, or just about any other type of input device that .NET can access. One or more blocks can be part of the pipeline, with each being responsible for a single operation. The following diagram illustrates two abstractions of dataflow pipelines:

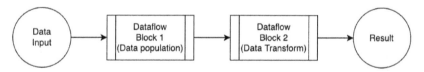

Example 1 - Load and transform data with dataflow pipeline

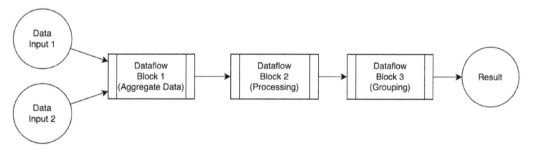

Example 2 - Aggregate, process, and group data with dataflow pipeline

Figure 7.1 – Dataflow pipeline examples

One real-world example you can consider is using a webcam to capture image frames. In a two-step flow, as shown in *Example 1*, consider the webcam as **Data Input**. **Dataflow Block 1** could perform some image processing to optimize the image appearance, while **Dataflow Block 2** will call an **Azure Cognitive Services** API to identify objects in each image. **Result** would contain a new .NET class for each input image containing the image binary data and properties that contain the identified objects within each image.

Next, let's learn about the types of blocks available in the dataflow library.

Types of dataflow blocks

There are nine predefined blocks in the dataflow library. These can be divided into three different categories. The first category is **buffering blocks**.

Buffering blocks

The purpose of **buffering blocks** is to buffer input data to be consumed. Buffering blocks are all propagator blocks, meaning they can be both a data source and target in a dataflow pipeline. There are three types of buffering blocks: `BufferBlock<T>`, `BroadcastBlock<T>`, and `WriteOnceBlock<T>`.

BufferBlock

`BufferBlock<T>` is an asynchronous queuing mechanism that implements a **first-in, first-out** (**FIFO**) queue of objects. `BufferBlock` can have multiple data sources and multiple targets configured. However, each message in a `BufferBlock` can only be delivered to one target block. The message is removed from the queue after it has been successfully delivered.

The following snippet pushes customer names into a `BufferBlock` and subsequently reads the first five names out to the console:

```
BufferBlock<string> customerBlock = new();
foreach (var customer in customers)
{
    await customerBlock.SendAsync(customer.Name);
}
for (int i = 0; i < 5; i++)
{
    Console.WriteLine(await customerBlock.ReceiveAsync());
}
// The code could display the following output:
//    Robert Jones
//    Jita Smith
```

```
//      Patty Xu
//      Sam Alford
//      Melissa Allen
```

BroadcastBlock

`BroadcastBlock<T>` is used similarly to `BufferBlock`, but it is intended to provide only the most recently posted message available to consumers. It can also be used to send the same value to many consumers. The message that's posted to a `BroadcastBlock` is not removed after it has been received by a consumer.

The following snippet will read the same alert message each time the `Receive` method is called:

```
var alertBlock = new BroadcastBlock<string>(null);
alertBlock.Post("Network is unavailable!");
for (int i = 0; i < 5; i++)
{
    Console.WriteLine(alertBlock.Receive());
}
```

WriteOnceBlock

As the name suggests, `WriteOnceBlock<T>` can only be written to once. After the first message has been received, all calls to `Post` or `SendAsync` will be ignored by the block. No exceptions will be thrown. The data is simply discarded.

The following example is similar to our `BufferBlock` snippet. However, because we're now using a `WriteOnceBlock`, only the first customer's name will be accepted by the block:

```
WriteOnceBlock<string> customerBlock = new();
foreach (var customer in customers)
{
    await customerBlock.SendAsync(customer.Name);
}
Console.WriteLine(await customerBlock.ReceiveAsync());
```

Execution blocks

Execution blocks are blocks that execute a delegate method for each message that's received. There are three types of execution blocks in the dataflow library. `ActionBlock<TInput>` is a **target** block, while `TransformBlock<TInput, TOuput>` and `TransformManyBlock<TInput, TOutput>` are both **propagator** blocks.

ActionBlock

ActionBlock is a block that accepts either Action<T> or Func<TInput, Task> as its constructor. An action on an input message is considered complete when the action returns or the task of Func completes. You can use an action for synchronous delegates or Func for async operations.

In this snippet, we will output customer names to the console with Console.WriteLine, which is provided in an Action, to the block:

```
var customerBlock = new ActionBlock<string>(name =>
    Console.WriteLine(name));
foreach (var customer in customers)
{
    await customerBlock.SendAsync(customer.Name);
}
customerBlock.Complete();
await customerBlock.Completion;
```

TransformBlock

TransformBlock<TInput, TOutput> is similar to ActionBlock. However, as a propagator block, it returns an output value for each message that's received. The two possible delegate signatures that can be provided to the TransformBlock constructor are Func<TInput, TOutput> for synchronous operations and Func<TInput, Task<TOutput>> for asynchronous operations.

The following example uses a TransformBlock that will convert a customer name into all capitals before the first five output values are retrieved to be displayed on the console:

```
var toUpperBlock = new TransformBlock<string, string>(name
        => name.ToUpper());
foreach (var customer in customers)
{
    toUpperBlock.Push(customer.Name);
}
for (int i = 0; i < 5; i++)
{
    Console.WriteLine(toUpperBlock.Receive());
}
```

TransformManyBlock

TransformManyBlock<TInput, TOutput> is similar to TransformBlock except that the block can return one or more values for every input value that's received. The possible delegate signatures for `TransformManyBlock` are `Func<TInput, IEnumerable<TOutput>>` and `Func<TInput, Task<IEnumerable<TOutput>>>` for synchronous and asynchronous operations, respectively.

In this snippet, we will pass one customer name to `TransformManyBlock`, which will return an enumerable containing the individual characters in the customer's name:

```
var nameCharactersBlock = new TransformManyBlock<string,
    char>(name => name.ToCharArray());
nameCharactersBlock.Post(customerName);
for (int i = 0; i < (customerName.Length; i++)
{
    Console.WriteLine(nameCharactersBlock.Receive());
}
```

Grouping blocks

Grouping blocks can combine objects from one or more sources. There are three types of grouping blocks. `BatchBlock<T>` is a propagator block, while `JoinBlock<T1, T2>` and `BatchedJoinBlock<T1, T2>` are both source blocks.

BatchBlock

`BatchBlock` accepts batches of data and produces arrays of output data. When creating a `BatchBlock`, you specify the input batch size. `BatchBlock` has a `Greedy` property in the `dataflowBlockOptions` optional constructor parameter that specifies the **greedy mode**:

- When `Greedy` is `true`, which is its default value, the block continues processing every input value as it is received and outputs an array as the batch size is reached.

- When `Greedy` is `false`, incoming messages can be paused while an array of the batch size is being created.

Greedy mode usually performs better, but if you are coordinating input from multiple sources, you may need to use **non-greedy mode**.

In this example, `BatchBlock` separates student names into classes with a maximum size of 12:

```
var studentBlock = new BatchBlock<string>(12);
// Assume studentList contains 20 students.
foreach (var student in studentList)
{
    studentBlock.Post(student.Name);
```

```
}
// Signal that we are done adding items.
studentBlock.Complete();
// Print the size of each class.
Console.WriteLine($"The number of students in class 1 is {
    studentBlock.Receive().Count()}.");  // 12 students
Console.WriteLine($"The number of students in class 2 is {
    studentBlock.Receive().Count()}.");  // 8 students
```

JoinBlock

JoinBlock has two signatures: JoinBlock<T1, T2> and JoinBlock<T1, T2, T3>. JoinBlock<T1, T2> has Target1 and Target2 properties to accept inputs and returns a Tuple<T1, T2> as each pair of targets is filled. JoinBlock<T1, T2, T3> has Target1, Target2, and Target3 properties and returns a Tuple<T1, T2, T3> as each set of targets is completed.

JoinBlock also has greedy and non-greedy modes, with greedy mode being the default behavior. When you switch to non-greedy mode, all input is postponed to targets that have already received input until a complete output set is populated and sent as output.

In this example, we will create a JoinBlock to combine a person's first name, last name, and age into the output tuple:

```
var joinBlock = new JoinBlock<string, string, int>();
joinBlock.Target1.Post("Sally");
joinBlock.Target1.Post("Raj");
joinBlock.Target2.Post("Jones");
joinBlock.Target2.Post("Gupta");
joinBlock.Target3.Post(7);
joinBlock.Target3.Post(23);
for (int i = 0; i < 2; i++)
{
    var data = joinBlock.Receive();
    if (data.Item3 < 18)
    {
        Console.WriteLine($"{data.Item1} {data.Item2} is a
            child.");
    }
    else
```

```
    {
        Console.WriteLine($"{data.Item1} {data.Item2} is
           an adult.");
    }
}
```

BatchedJoinBlock

A `BatchedJoinBlock` is like a `JoinBlock` except the tuple in the output contains `IList` items of the size of the batch specified in the constructor: `Tuple(IList(T1), IList(T2))` or `Tuple(IList(T1), IList(T2), IList(T3))`. The batching concept is the same as it is for `BatchBlock`.

As an exercise, try to build on the `JoinBlock` example to add more people to the list, divide them into batches of four, and output the name of the oldest person in each batch.

Now that we have explored examples of all of the available dataflow blocks, let's get into some real-world dataflow examples. In the next section, we will use some dataflow blocks to create a producer/consumer implementation.

Implementing the producer/consumer pattern

The blocks in the TPL Dataflow library provide a fantastic platform for implementing the **producer/consumer pattern**. If you are not familiar with this design pattern, it involves two operations and a queue of work. The **producer** is the first operation. It is responsible for filling the queue with data or units of work. The **consumer** is responsible for taking items from the queue and acting on them in some way. There can be one or more producers and one or more consumers in the system. You can change the number of producers or consumers, depending on which part of the process is the bottleneck.

> **Real-World Scenario Example**
>
> To relate the producer/consumer pattern to a *real-world scenario*, think about preparing gifts for a holiday gathering. You and a partner are working together to prepare the gifts. You are fetching and staging the gifts to be wrapped. You are the *producer*. Your partner is taking items from your queue and wrapping each gift. They are the *consumer*. If the queue starts to get backed up, you can find another friend (or consumer) to help with the wrapping and increase the overall throughput. If, on the other hand, you are taking too much time to find each gift to be wrapped, you can add another producer to help find them and fill the queue. This will keep the consumers busy and increase the efficiency of the process.

In our .NET producer/consumer example, we are going to build a simple WPF application that fetches blog posts from multiple RSS feeds and displays them in a single `ListView` control. Each row in the list will include the blog post's date, categories, and an HTML summary of the post's content. The producers in the application will fetch posts from an RSS feed and add a `SyndicationItem` to the queue for each blog post. We will get posts from three blogs and create a producer for each.

The consumers will take a `SyndicationItem` from the queue and use an `ActionBlock` delegate to create a `BlogPost` object for each `SyndicationItem`. We will create three consumers to keep up with the items that have been queued by our three producers. When the process completes, the list of `BlogPost` objects will be set as `ItemSource` for `ListView`. Let's get started:

1. Start by creating a new WPF project with .NET 6. Name the project `ProducerConsumerRssFeeds`.

2. Open **NuGet Package Manager** for the solution, search for **Syndication** on the **Install** tab, and add the **System.ServiceModel.Syndication** package to the project. This package will make it simple to fetch data from any RSS feed.

3. Add a new class to the project named `BlogPost`. This will be our model object for each blog post to be displayed in `ListView`. Add the following properties to the new class:

    ```
    public class BlogPost
    {
        public string PostDate { get; set; } = "";
        public string? Categories { get; set; }
        public string? PostContent { get; set; }
    }
    ```

4. Now, it's time to create a service class to fetch the blog posts for a given RSS feed URL. Add a new class named `RssFeedService` to the project and add a method named `GetFeedItems` to the class:

    ```
    using System.Collections.Generic;
    using System.ServiceModel.Syndication;
    using System.Xml;
    ...
    public static IEnumerable<SyndicationItem>
        GetFeedItems(string feedUrl)
    {
        using var xmlReader = XmlReader.Create(feedUrl);
        SyndicationFeed rssFeed = SyndicationFeed.Load
            (xmlReader);
        return rssFeed.Items;
    }
    ```

The static `SyndicationFeed.Load` method uses `XmlReader` to fetch the XML from the provided `feedUrl` and transform it into `IEnumerable<SyndicationItem>` to return from the method.

5. Next, create a new class named `FeedAggregator`. This class will contain the producer/ consumer logic that calls `GetFeedItems` for each blog and transforms the feed data for each blog post so that it can be displayed in the UI. The three blogs that we are aggregating are as follows:

- The .NET blog

- The Windows blog

- The Microsoft 365 blog

The first step with `FeedAggregator` is creating a producer method named `ProduceFeedItems` and a parent method named `QuseueAllFeeds` that will start three instances of the producer method:

```
private async Task QueueAllFeeds(BufferBlock
    <SyndicationItem> itemQueue)
{
    Task feedTask1 = ProduceFeedItems(itemQueue,
        "https://devblogs.microsoft.com/dotnet/feed/");
    Task feedTask2 = ProduceFeedItems(itemQueue,
        "https://blogs.windows.com/feed");
    Task feedTask3 = ProduceFeedItems(itemQueue,
        "https://www.microsoft.com/microsoft-
            365/blog/feed/");
    await Task.WhenAll(feedTask1, feedTask2,
        feedTask3);
    itemQueue.Complete();
}
private async Task ProduceFeedItems
    (BufferBlock<SyndicationItem> itemQueue, string
        feedUrl)
{
    IEnumerable<SyndicationItem> items =
        RssFeedService.GetFeedItems(feedUrl);
    foreach (SyndicationItem item in items)
    {
```

```
        await itemQueue.SendAsync(item);
    }
}
```

We are using `BufferBlock<SyndicationItem>` as our queue. Every producer calls `GetFeedItems` and adds each `SyndicationItem` that's returned to `BufferBlock`. The `QueueAllFeeds` method uses `Task.WhenAll` to wait for all of the producers to finish adding items to the queue. Then, it signals to `BufferBlock` that all the producers are done by calling `itemQueue.Complete()`.

6. Next, we will create our consumer method. This method, named `ConsumeFeedItem`, will be responsible for taking a `SyndicationItem` provided by `BufferBlock` and converting it into a `BlogPost` object. Each `BlogPost` will be added to `ConcurrentBag<BlogPost>`. We're using a thread-safe collection here because there will be multiple consumers adding output to the list:

```
private void ConsumeFeedItem(SyndicationItem nextItem,
    ConcurrentBag<BlogPost> posts)
{
    if (nextItem != null && nextItem.Summary != null)
    {
        BlogPost newPost = new();
        newPost.PostContent = nextItem.Summary.Text
            .ToString();
        newPost.PostDate = nextItem.PublishDate
            .ToLocalTime().ToString("g");
        if (nextItem.Categories != null)
        {
            newPost.Categories = string.Join(",",
                nextItem.Categories.Select(c =>
                    c.Name));
        }
        posts.Add(newPost);
    }
}
```

7. Now, it's time to tie the producer/consumer logic together. Create a method named `GetAllMicrosoftBlogPosts`:

```
public async Task<IEnumerable<BlogPost>>
    GetAllMicrosoftBlogPosts()
```

```
{
        var posts = new ConcurrentBag<BlogPost>();
        // Create queue of source posts
        BufferBlock<SyndicationItem> itemQueue = new(new
            DataflowBlockOptions { BoundedCapacity =
                10 });
        // Create and link consumers
        var consumerOptions = new Execution
            DataflowBlockOptions { BoundedCapacity = 1 };
        var consumerA = new ActionBlock<SyndicationItem>
            ((i) => ConsumeFeedItem(i, posts),
                consumerOptions);
        var consumerB = new ActionBlock<SyndicationItem>
            ((i) => ConsumeFeedItem(i, posts),
                consumerOptions);
        var consumerC = new ActionBlock<SyndicationItem>
            ((i) => ConsumeFeedItem(i, posts),
                consumerOptions);
        var linkOptions = new DataflowLinkOptions {
            PropagateCompletion = true, };
        itemQueue.LinkTo(consumerA, linkOptions);
        itemQueue.LinkTo(consumerB, linkOptions);
        itemQueue.LinkTo(consumerC, linkOptions);
        // Start producers
        Task producers = QueueAllFeeds(itemQueue);
        // Wait for producers and consumers to complete
        await Task.WhenAll(producers, consumerA.Completion,
            consumerB.Completion, consumerC.Completion);
        return posts;
}
```

I.   The method starts by creating a ConcurrentBag<BlogPost> to aggregate the final list of posts for the UI. Then, it creates the itemQueue object with a BoundedCapacity of 10. This bounded capacity means that no more than 10 items can be enqueued at any time. Once the queue reaches 10, all the producers must wait for the consumers to dequeue some items. This can slow the performance of the process, but it prevents potential out-of-memory issues in production code. Our sample is not in any danger of running out of memory when processing posts from three blogs, but you can see how to use BoundedCapacity when it is needed in your applications. You can create the queue with no BoundedCapacity like this:

```
BufferBlock<SyndicationItem> itemQueue = new();
```

II.   The next part of the method creates three consumers that use ActionBlock<SyndicationItem> with ConsumeFeedItem as the provided delegate. Each consumer is linked to the queue with the LinkTo method. Setting BoundedCapacity of the consumers to 1 tells the producers to move on to the next consumer if the current one is already busy processing an item.

III.   Once the links have been established, we can start the producers by calling QueueAllFeeds. Then, we must await the producers and the Completion object of each consumer ActionBlock. By linking the completion of the producers and consumers, we don't need to explicitly await the Completion object of the consumers:

```
var linkOptions = new DataflowLinkOptions {
    PropagateCompletion = true, };
```

8. The next step is to create some UI controls to display the information to our users. Open the MainWindow.xaml file and replace the existing Grid with the following markup:

```
<Grid>
    <ListView x:Name="mainListView">
        <ListView.ItemTemplate>
            <DataTemplate>
                <Grid>
                    <Grid.ColumnDefinitions>
                        <ColumnDefinition
                            Width="150"/>
                        <ColumnDefinition
                            Width="300"/>
                        <ColumnDefinition
                            Width="500"/>
                    </Grid.ColumnDefinitions>
                    <TextBlock Grid.Column="0"
                        Text="{Binding PostDate}"
                            Margin="3"/>
                    <TextBox IsReadOnly="True"
```

```
                            Grid.Column="1"
                        Text="{Binding Categories}"
                            Margin="3"
                                TextWrapping="Wrap"/>
                    <TextBox IsReadOnly="True"
                        Grid.Column="2"
                          Text="{Binding PostContent}"
                            Margin="3"/>
                </Grid>
            </DataTemplate>
        </ListView.ItemTemplate>
    </ListView>
</Grid>
```

Explaining the details of WPF, XAML, and data binding are outside the scope of this book. If you would like to learn more about WPF, check out *Mastering Windows Presentation Foundation*, by Sheridan Yeun: `https://www.packtpub.com/product/mastering-windows-presentation-foundation-second-edition/9781838643416`. What this markup does is create a new `ListView` control with a `DataTemplate` to define the structure of each list item in the control. For each item, we are defining either a `TextBlock` or `TextBox` to hold the values for each `BlogPost` object in the list.

9. The last thing we must do is call the `GetAllMicrosoftBlogPosts` method and populate the UI. Open `MainWindow.xaml.cs` and add the following code:

```
public MainWindow()
{
    InitializeComponent();
    Loaded += MainWindow_Loaded;
}
private async void MainWindow_Loaded(object sender,
    RoutedEventArgs e)
{
    FeedAggregator aggregator = new();
    var items = await aggregator
        .GetAllMicrosoftBlogPosts();
    mainListView.ItemsSource = items;
}
```

After `MainWindow` has loaded, the items that have been returned from `GetAllMicrosoftBlogPosts` are set as `mainListView.ItemsSource`. This will allow the data to bind to the elements in `DataTemplate`, which we defined in the XAML.

10. Now, run the project and see how things look:

Figure 7.2 – Running the ProducerConsumerRssFeeds WPF application for the first time

As you can see, the list displays 10 blog post summaries from each of the Microsoft blogs. This is the default maximum number of items that can be returned by Microsoft's blogs.

You can try experimenting by increasing or decreasing the number of producers and consumers in the project. Does adding more consumers speed up the process? Try adding some of your favorite blogs' feeds to the list of producers and see what happens.

> **Note**
>
> You may have noticed that the content summary that's returned by the RSS feeds contains HTML, and we are just rendering it as plain text in a `TextBox` control. If you would like to use a `RichTextBox` that renders the HTML properly, take a look at this sample project on CodeProject that uses a WPF **Behavior** to render HTML in a `RichTextBox`: `https://www.codeproject.com/articles/1097390/displaying-html-in-a-wpf-richtextbox`.

In the next section, we will create another example that uses different types of dataflow blocks to create a data pipeline.

Creating a data pipeline with multiple blocks

One of the biggest advantages of using dataflow blocks is the ability to link them and create a complete workflow or data pipeline. In the previous section, we saw how this linking worked between producer and consumer blocks. In this section, we will create a console application with a pipeline of five dataflow blocks all linked together to complete a series of tasks. We will leverage `TransformBlock`, `TransformManyBlock`, and `ActionBlock` to take an RSS feed and output a list of categories that are unique across all blog posts in the feed. Follow these steps:

1. Start by creating a new .NET 6 console application in Visual Studio named `OutputBlogCategories`.

2. Add the **System.ComponentModel.Syndication** NuGet package that we used in the previous example.

3. Add the same `RssFeedService` class from the previous example. You can right-click on the project in **Solution Explorer** and select **Add | Existing Item** or you can create a new class named `RssFeedService` and copy/paste the same code we used in the previous example.

4. Add a new class named `FeedCategoryTransformer` to the project and create a method named `GetCategoriesForFeed`:

   ```
   public static async Task GetCategoriesForFeed(string
       url)
   {
   }
   ```

5. Over the next few steps, we will create the implementation for the `GetCategoriesForFeed` method. First, create a `TransformBlock` named `downloadFeed` that accepts `url` as a string and returns `IEnumerable<SyndicationItem>` from the `GetFeedItems` method:

   ```
   // Downloads the requested blog posts.
   var downloadFeed = new TransformBlock<string,
       IEnumerable<SyndicationItem>>(url =>
   {
       Console.WriteLine("Fetching feed from '{0}'...",
           url);
       return RssFeedService.GetFeedItems(url);
   });
   ```

6.   Next, create a `TransformBlock` that accepts `IEnumerable<SyndicationItem>` and returns `List<SyndicationCategory>`. This block will fetch the complete list of categories from every blog post and return them as a single list:

```
// Aggregates the categories from all the posts.
var createCategoryList = new TransformBlock
    <IEnumerable<SyndicationItem>, List
        <SyndicationCategory>>(items =>
{
    Console.WriteLine("Getting category list...");
    var result = new List<SyndicationCategory>();
    foreach (var item in items)
    {
        result.AddRange(item.Categories);
    }
    return result;
});
```

7.   Now, create another `TransformBlock`. This block will accept `List<SyndicationCategory>` from the previous block, remove all duplicates, and return the filtered `List<SyndicationCategory>`:

```
// Removes duplicates.
var deDupList = new TransformBlock<List
    <SyndicationCategory>, List<SyndicationCategory>>
        (categories =>
{
    Console.WriteLine("De-duplicating category
        list...");
    var categoryComparer = new CategoryComparer();
    return categories.Distinct(categoryComparer)
        .ToList();
});
```

To use the LINQ Distinct extension method on a complex object such as `SyndicationCategory`, a custom comparer that implements `IEqualityComparer<T>` is required. You can get the complete source for `CategoryComparer` from this chapter's GitHub repository: `https://github.com/PacktPublishing/Parallel-Programming-and-Concurrency-with-C-sharp-10-and-.NET-6/tree/main/chapter07`.

8. Next, create a `TransformManyBlock` named `createCategoryString`. This block will accept the de-duplicated `List<SyndicationCategory>` and return a string for each `Name` property of the categories. So, the block is invoked once for the entire list, but it will, in turn, invoke the next block in the flow once for every item in the list:

```
// Gets the category names from the list of category
    objects.
var createCategoryString = new TransformManyBlock
    <List<SyndicationCategory>, string>(categories =>
{
    Console.WriteLine("Extracting category names...");
    return categories.Select(c => c.Name);
});
```

9. The final block is an `ActionBlock` named `printCategoryInCaps`. This block will output each category name to the console in all caps using `ToUpper`:

```
// Prints the upper-cased unique categories to the
    console.
var printCategoryInCaps = new ActionBlock<string>
    (categoryName =>
{
    Console.WriteLine($"Found CATEGORY
        {categoryName.ToUpper()}");
});
```

10. Now that the dataflow blocks have been configured, it's time to link them. Create a `DataflowLinkOptions` that will propagate the completion of each block. Then, use the `LinkTo` method to link each block in the chain to the next one:

```
var linkOptions = new DataflowLinkOptions {
    PropagateCompletion = true };
downloadFeed.LinkTo(createCategoryList, linkOptions);
createCategoryList.LinkTo(deDupList, linkOptions);
deDupList.LinkTo(createCategoryString, linkOptions);
createCategoryString.LinkTo(printCategoryInCaps,
    linkOptions);
```

11. The last few steps of creating the `GetCategoriesForFeed` method involve sending `url` to the first block, marking it as `Complete`, and waiting for the last block in the chain:

```
await downloadFeed.SendAsync(url);
downloadFeed.Complete();
await printCategoryInCaps.Completion;
```

12. Now, open `Program.cs` and update the code so that it calls `GetCategoriesForFeed`, providing the URL for the Windows blog RSS feed:

```
using OutputBlogCategories;
Console.WriteLine("Hello, World!");
await FeedCategoryTransformer.GetCategoriesForFeed
    ("https://blogs.windows.com/feed");
Console.ReadLine();
```

13. Run the program and examine the list of categories in the output:

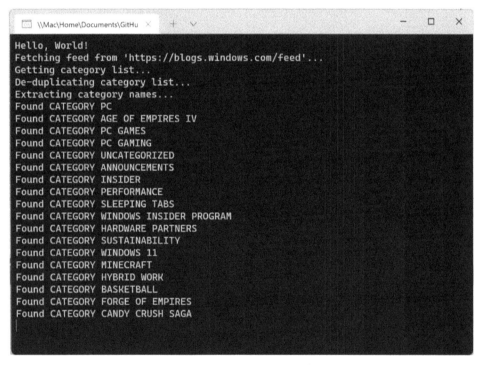

Figure 7.3 – Displaying a deduplicated list of categories from the Windows blog feed

Now that you understand how to create a data pipeline with a series of dataflow blocks, we will look at an example of combining data from multiple sources with a `JoinBlock`.

Manipulating data from multiple data sources

A `JoinBlock` can be configured to receive different data types from two or three data sources. As each set of data types is completed, the block is completed with a `Tuple` containing all three object types to be acted upon. In this example, we will create a `JoinBlock` that accepts a `string` and `int` pair and passes `Tuple(string, int)` along to an `ActionBlock`, which outputs their values to the console. Follow these steps:

1. Start by creating a new console application in Visual Studio

2. Add a new class named `DataJoiner` to the project and add a static method to the class named `JoinData`:

```
public static void JoinData()
{

}
```

3. Add the following code to create two `BufferBlock` objects, a `JoinBlock<string, int>`, and an `ActionBlock<Tuple<string, int>>`:

```
var stringQueue = new BufferBlock<string>();
var integerQueue = new BufferBlock<int>();
var joinStringsAndIntegers = new JoinBlock<string,
    int>(
    new GroupingDataflowBlockOptions
    {
        Greedy = false
    });
var stringIntegerAction = new ActionBlock
    <Tuple<string, int>>(data =>
{
    Console.WriteLine($"String received:
        {data.Item1}");
    Console.WriteLine($"Integer received:
        {data.Item2}");
});
```

Setting the block to non-greedy mode means it will wait for an item of each type before executing the block.

4. Now, create the links between the blocks:

```
stringQueue.LinkTo(joinStringsAndIntegers.Target1);
integerQueue.LinkTo(joinStringsAndIntegers.Target2);
joinStringsAndIntegers.LinkTo(stringIntegerAction);
```

5. Next, push some data to the two `BufferBlock` objects, wait for a second, and then mark them both as complete:

```
stringQueue.Post("one");
stringQueue.Post("two");
stringQueue.Post("three");
integerQueue.Post(1);
integerQueue.Post(2);
integerQueue.Post(3);
stringQueue.Complete();
integerQueue.Complete();
Thread.Sleep(1000);
Console.WriteLine("Complete");
```

6. Add the following code to `Program.cs` to run the example code:

```
using JoinBlockExample;
DataJoiner.JoinData();
Console.ReadLine();
```

7. Finally, run the application and examine the output. You will see that `ActionBlock` outputs a `string` and `integer` pair for each set of values provided:

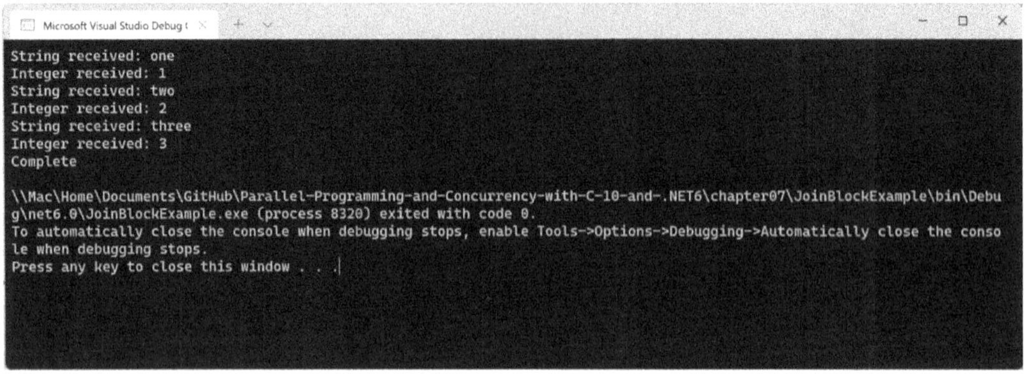

Figure 7.4 – Running the JoinBlockExample console application

That's all there is to using the `JoinBlock` dataflow block. Try making some changes on your own, such as changing the `Greedy` option or the order in which data is added to each `BufferBlock`. How does that impact the output?

Before we finish up this chapter, let's review everything we've learned.

Summary

In this chapter, we learned all about the various blocks in the TPL Dataflow library. We started by learning a little about each block type and providing a brief code snippet for each. Next, we created a practical example that implemented the producer/consumer pattern to fetch blog data from three different Microsoft blogs. We also examined `TransformBlock`, `TransformManyBlock`, and `JoinBlock` more closely in .NET console applications. You should now feel confident in your ability to use some of the dataflow blocks in your applications to automate some complex data workflows.

If you would like some additional reading about the TPL Dataflow library, you can download *Introduction to TPL Dataflow* from the Microsoft Download Center: `https://www.microsoft.com/en-us/download/details.aspx?id=14782`.

In the next chapter, *Chapter 8*, we will take a closer look at the collections in the `System.Collections.Concurrent` namespace. We will also discover some practical uses of PLINQ in modern .NET applications.

Questions

Answer the following questions to test your knowledge of this chapter:

1. What type of data flow block aggregates data from two or three data sources?
2. What type of block is a `BufferBlock`?
3. What type of block is populated by a producer in the producer/consumer pattern?
4. What method links the completion of two blocks?
5. What method is called to signal that our code is done adding data to a source block?
6. What is the async equivalent of calling `Post()`?
7. What is the async equivalent of calling `Receive()`?

8

Parallel Data Structures and Parallel LINQ

.NET provides many useful features and data constructs for developers who are introducing parallelism to their projects. This chapter will explore these features, including **concurrent collections**, the `SpinLock<T>` **synchronization primitive**, and **Parallel LINQ (PLINQ)**. These features can improve the performance of your applications while maintaining safe threading practices.

Most .NET developers are familiar with LINQ frameworks, including LINQ to Objects, LINQ to SQL, and LINQ to XML. There are even open source .NET LINQ libraries, such as LINQ to Twitter (`https://github.com/JoeMayo/LinqToTwitter`). We will take those LINQ skills and leverage them in parallel programming with PLINQ. Every LINQ developer can be a PLINQ developer after reading this chapter. Read ahead for some useful examples of working with PLINQ in C#.

In this chapter, you will learn about the following:

- Introducing PLINQ
- Converting LINQ queries to PLINQ
- Preserving data order and merging data with PLINQ
- Data structures for parallel programming in .NET

By the end of this chapter, you will have a new appreciation for LINQ when it comes to parallel programming.

Technical requirements

To follow along with the examples in this chapter, the following software is recommended for Windows developers:

- Visual Studio 2022 version 17.0 or later.

- .NET 6.

- To complete the WPF sample, you will need to install the .NET desktop development workload for Visual Studio.

While these are recommended, if you have .NET 6 installed, you can use your preferred editor. For example, Visual Studio 2022 for Mac on macOS 10.13 or later, JetBrains Rider, or Visual Studio Code will work just as well.

All the code examples for this chapter can be found on GitHub at https://github.com/ PacktPublishing/Parallel-Programming-and-Concurrency-with-C-sharp-10-and-.NET-6/tree/main/chapter08.

Let's get started by discussing LINQ, PLINQ, and why the query language can be a great way to improve your parallel programming with C#.

Introducing PLINQ

PLINQ is a set of .NET extensions for LINQ that allow part of the LINQ query to execute in parallel by leveraging the thread pool. The PLINQ implementation provides parallel versions of all of the available LINQ query operations.

Like LINQ queries, PLINQ queries offer deferred execution. This means that the objects are not queried until they need to be enumerated. If you aren't familiar with LINQ's deferred execution, we will look at a simple example to illustrate the concept. Consider these two LINQ queries:

```
internal void QueryCities(List<string> cities)
{
    // Query is executed with ToList call
    List<string> citiesWithS = cities.Where(s =>
        s.StartsWith('S')).ToList();
    // Query is not executed here
    IEnumerable<string> citiesWithT = cities.Where(s =>
        s.StartsWith('T'));
    // Query is executed here when enumerating
    foreach (string city in citiesWithT)
    {
```

```
        // Do something with citiesWithT
    }
}
```

In the example, the LINQ query that populates `citiesWithS` is executed immediately because of the call to `ToList()`. The second query that populates `citiesWithT` is not immediately executed. The execution is deferred until the `IEnumerable` values are required. The `citiesWithT` values are not required until we iterate over them in the `foreach` loop. The same principle holds true for PLINQ queries.

> **Note**
>
> If you are unfamiliar with LINQ concepts or the LINQ method syntax, the book *C# 10 and .NET 6 – Modern Cross-Platform Development – Sixth Edition* by *Mark J. Price* has a chapter dedicated to explaining LINQ syntax and several of its implementations. It is an excellent book for every .NET developer. You can find out more about the book here: `https://subscription.` `packtpub.com/product/mobile/9781801077361`.

PLINQ is similar to LINQ in other ways, too. You can create PLINQ queries on any collection that implements `IEnumerable` or `IEnumerable<T>`. You can use all of the familiar LINQ operations such as `Where`, `FirstOrDefault`, `Select`, and so on. The primary difference is that PLINQ attempts to leverage the power of parallel programming by part or all of a query across multiple threads. Internally, PLINQ partitions the in-memory data into multiple segments and performs the query on each segment in parallel.

There are several factors that impact the performance gained by using PLINQ. Let's explore those next.

PLINQ and performance

When deciding which LINQ queries are good candidates to leverage the power of PLINQ, you must consider a number of factors. The primary factor to consider is whether the magnitude or complexity of the work to be performed is great enough to offset the overhead of threading. You should be operating on a large dataset and be performing an expensive operation on each item in the collection. The LINQ example that checked the first letter of a string is not a very good candidate for PLINQ, especially if the source collection only contains a handful of items.

Another factor in the performance to potentially be gained with PLINQ is the number of cores available on the system where the queries will be running. The greater the number of cores that PLINQ can leverage, the better the potential gain. PLINQ can break down a large dataset into more units of work to be executed in parallel with many cores at its disposal.

Ordering and grouping data can incur a larger amount of overhead than it would in a traditional LINQ query. The PLINQ data is segmented, but grouping and ordering must be performed across the entire collection. PLINQ is best suited for queries where the data sequence is not important.

We will discuss some other factors that impact query performance in the *Preserving data order and merging data with PLINQ* section. Now, let's start creating our first PLINQ queries.

Creating a PLINQ query

The majority of the functionality of PLINQ is exposed through members of the `System.Linq.ParallelEnumerable` class. This class contains implementations of all of the LINQ operators that are available to in-memory object queries. There are some additional operators in this class that are specific to PLINQ queries. The two most important operators to understand are `AsParallel` and `AsSequential`. The `AsParallel` operator indicates that all subsequent LINQ operations should be attempted to be performed in parallel. In contrast, the `AsSequential` operator indicates to PLINQ that the LINQ operations that follow it should be performed in sequence.

Let's look at an example that uses both of these PLINQ operators. Our query will be operating on `List<Person>` with the following definition:

```
internal class Person
{
    public string FirstName { get; set; } = "";
    public string LastName { get; set; } = "";
    public int Age { get; set; }
}
```

Let's consider that we are working with a dataset of thousands or even millions of people. We want to leverage PLINQ to query only the adults aged 18 or older from the data and then group them by their last name. We want to execute only the `Where` clause of the query in parallel. The `GroupBy` operation will be performed sequentially. This method will do exactly that:

```
internal void QueryAndGroupPeople(List<Person> people)
{
    var results = people.AsParallel().Where(p => p.Age > 17)
        .AsSequential().GroupBy(p => p.LastName);
    foreach (var group in results)
    {
        Console.WriteLine($"Last name {group.Key} has
            {group.Count()} people.");
```

```
        }
        // Sample output:
        // Last name Jones has 4220 people.
        // Last name Xu has 3434 people.
        // Last name Patel has 4798 people.
        // Last name Smith has 3051 people.
        // Last name Sanchez has 3811 people.
        // ...
    }
```

The GroupBy LINQ method will return IEnumerable<IGrouping<string, Person>> with each IGrouping<string, Person> instance containing all of the people with the same LastName. Whether or not this GroupBy operation would be faster to run in parallel or sequentially depends on the makeup of the data. You should always test your application to determine whether introducing parallelism is improving the performance when working with production data. We will cover ways to performance-test your code in *Chapter 10*.

Next, let's look at how PLINQ queries can be written with the **method syntax** that we have used thus far or by using LINQ **query syntax**.

Query syntax versus method syntax

LINQ queries can be coded either by using method syntax or query syntax. Method syntax is where you string multiple methods together to build a query. This is what we have been doing throughout this section. Query syntax is slightly different, and it is akin to T-SQL query syntax. Let's examine the same PLINQ query written both ways.

Here is a simple PLINQ query to return only adults from a list of people written with method syntax:

```
var peopleQuery1 = people.AsParallel().Where(p => p.Age > 17);
```

Here is the exact same PLINQ query written with query syntax:

```
var peopleQuery2 = from person in people.AsParallel()
                   where person.Age > 17
                   select person;
```

You should use whichever syntax you prefer. Throughout the rest of this chapter, we will be using method syntax for the examples.

In the next section, we will continue to explore the operations available in PLINQ and create some parallel versions of LINQ queries.

Converting LINQ queries to PLINQ

In this section, we will look at some additional PLINQ operators and show you how you can leverage them to turn existing LINQ queries into PLINQ queries. Your existing queries may have requirements for preserving the order of data. Perhaps your existing code doesn't use LINQ at all. There could be an opportunity there to convert some logic in `foreach` loops into PLINQ operations.

The way to convert a LINQ query to a PLINQ query is by inserting an `AsParallel()` statement into the query, as we did in the previous section. Everything that follows `AsParallel()` will run in parallel until an `AsSequential()` statement is encountered.

If your queries require that the original order of objects be preserved, you can include an `AsOrdered()` statement:

```
var results = people.AsParallel().AsOrdered()
    .Where(p => p.LastName.StartsWith("H"));
```

However, this will not be as performant as queries that do not preserve the sequence of data. To explicitly tell PLINQ to not preserve data order, use the `AsUnordered()` statement:

```
var results = people.AsParallel().AsUnordered()
    .Where(p => p.LastName.StartsWith("H"));
```

The unordered version of the query will perform much better if the order of your data is not important; you should never use `AsOrdered()` with PLINQ.

Let's consider another example. We will start with a method that iterates over a list of people with a `foreach` loop and calls a method named `ProcessVoterActions` for each person aged 18 or older. We're going to assume that this method is processor-intensive and also uses some I/O to save the voter information in a database. Here is the starting code:

```
internal void ProcessAdultsWhoVote(List<Person> people)
{
    foreach (var person in people)
    {
        if (person.Age < 18) continue;
        ProcessVoterActions(person);
    }
}
```

```
private void ProcessVoterActions (Person adult)
{
    // Add adult to a voter database and process their
        data.
}
```

This will not leverage parallel processing at all. We could improve on this by using LINQ to filter out the children under 18 and then call `ProcessVoterActions` with a `Parallel.ForEach` loop:

```
internal void ProcessAdultsWhoVoteInParallel (List<Person>
    people)
{
    var adults = people.Where(p => p.Age > 17);
    Parallel.ForEach(adults, ProcessVoterActions);
}
```

This will certainly improve the performance if `ProcessVoterActions` takes some time to run for each person. However, with PLINQ, we can improve the performance even further:

```
internal void ProcessAdultsWhoVoteWithPlinq (List<Person>
    people)
{
    var adults = people.AsParallel().Where(p => p.Age > 17);
    adults.ForAll(ProcessVoterActions);
}
```

Now, the `Where` query will run in parallel. This will certainly help performance if we expect to have thousands or millions of objects in the `people` collection. The `ForAll` extension method is another PLINQ operation that runs in parallel. It is meant to be used to perform an operation in parallel on each object in the query results.

The performance of `ForAll` will also be superior to the `Parallel.ForEach` operation in the previous example. One difference is the deferred execution of PLINQ. These calls to `ProcessVoterActions` will not be performed until the `IEnumerable` result is iterated over. The other advantage is the same advantage over performing a standard `foreach` loop with `IEnumerable` after completing a PLINQ query on your data. The data must be merged back from the multiple threads before it can be enumerated by either `foreach` or `Parallel.ForEach`. With a `ForAll` operation, the data can remain segmented by PLINQ and merged once at the end. This diagram illustrates the difference between `Parallel.ForEach` and `ForAll`:

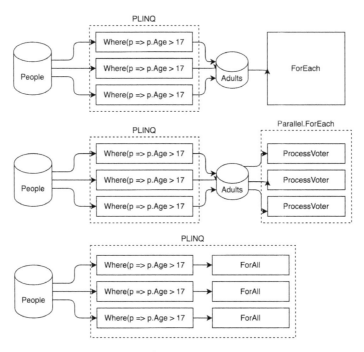

Figure 8.1 – Advantages of PLINQ, data segmentation, and ForAll

Before we explore more details about data order and merging data, let's discuss how to handle exceptions when working with PLINQ.

Handling exceptions with PLINQ queries

Implementing good exception handling in your .NET projects is important. It's one of the fundamental practices of software development. When working with parallel programming in general, exception handling can be more complicated. This is also true with PLINQ. When any exception is unhandled within a parallel operation inside a PLINQ query, the query will throw an exception of type `AggregateException`. So, at the very minimum, all of your PLINQ queries should run within a `try/catch` block that catches the `AggregateException` exception type.

Let's take our PLINQ `ForAll` example with `ProcessVoterActions` and add some exception handling:

1. We're going to run this example in a .NET console application, so create a new project in Visual Studio and add a class named `Person`:

    ```
    internal class Person
    {
        public string FirstName { get; set; } = "";
    ```

```
        public string LastName { get; set; } = "";
        public int Age { get; set; }
    }
```

2. Next, add a new class named `PlinqExceptionsExample`.

3. Now add a private method to `PlinqExceptionsExample` named `ProcessVoterActions`. We're going to throw `ArgumentException` for any person older than `120`:

```
    private void ProcessVoterActions(Person adult)
    {
        if (adult.Age > 120)
        {
            throw new ArgumentException("This person is
                too old!", nameof(adult));
        }
        // Add adult to a voter database and process their
    data.
    }
```

4. Next, add the `ProcessAdultsWhoVoteWithPlinq` method:

```
    internal void ProcessAdultsWhoVoteWithPlinq
        (List<Person> people)
    {
        try
        {
            var adults = people.AsParallel().Where(p =>
                p.Age > 17);
            adults.ForAll(ProcessVoterActions);
        }
        catch (AggregateException ae)
        {
            foreach (var ex in ae.InnerExceptions)
            {
                Console.WriteLine($"Exception encountered
                    while processing voters. Message:
                        {ex.Message}");
```

```
                }
            }
        }
```

This method's logic remains the same. It's filtering out the children with a PLINQ Where clause and calling ProcessVoterActions as a delegate to ForAll.

> **Note**
>
> If you are following along with the sample code on GitHub for this chapter (https://github.com/PacktPublishing/Parallel-Programming-and-Concurrency-with-C-sharp-10-and-.NET-6/tree/main/chapter08/LINQandPLINQsnippets), you will need to uncomment the lines of code in *Step 5*. You should also comment out the lines in the Main method that follow those lines to prevent other samples from executing.

5. Finally, open Program.cs and add some code to create an instance of List<Person> in an inline function called GetPeople. It can contain as many people as you like, but at least one of them needs to have an age greater than 120. Call ProcessAdultsWhoVoteWithPlinq, passing the data from GetPeople:

```
using LINQandPLINQsnippets;
var exceptionExample = new PlinqExceptionsExample();
exceptionExample.ProcessAdultsWhoVoteWithPlinq
    (GetPeople());
Console.ReadLine();

static List<Person> GetPeople()
{
    return new List<Person>
    {
        new Person { FirstName = "Bob", LastName =
            "Jones", Age = 23 },
        new Person { FirstName = "Sally", LastName =
            "Shen", Age = 2 },
        new Person { FirstName = "Joe", LastName =
            "Smith", Age = 45 },
        new Person { FirstName = "Lisa", LastName =
            "Samson", Age = 98 },
        new Person { FirstName = "Norman", LastName =
            "Patterson", Age = 121 },
```

```
            new Person { FirstName = "Steve", LastName =
                "Gates", Age = 40 },
            new Person { FirstName = "Richard", LastName =
                "Ng", Age = 18 }
        };
    }
```

6. Now, run the program and observe the console output. If Visual Studio breaks at the exception, just click **Continue**:

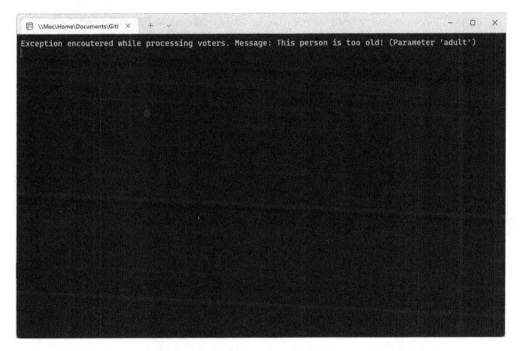

Figure 8.2 – Receiving an exception in the console

The problem with handling the exception from outside the PLINQ query is that the entire query is stopped. It isn't able to run to completion. If you have an exception that shouldn't stop the entire process, you should handle it from within the code inside the query and continue processing the remaining items.

If you handle exceptions inside ProcessVoterActions, you have a chance to process them gracefully and continue.

Next, we are going to explore some examples of how to preserve the order of your data and handle different options for merging segments back together.

Preserving data order and merging data with PLINQ

When fine-tuning PLINQ queries for your applications, there are some extension methods that impact the sequencing of data that you can leverage. Preserving the original order of your items may be something that is required. We have touched on the `AsOrdered` method in this chapter, and we will experiment with it in this section. When PLINQ operations have been completed and items are returned as part of the final enumeration, the data is merged from the segments that were created to operate on multiple threads. The merge behavior can be controlled by setting `ParallelMergeOptions` with the `WithMergeOptions` extension method. We will discuss the behavior of the three available merge options provided.

Let's get started by creating some samples with the `AsOrdered` and `AsUnordered` extension methods.

PLINQ data order samples

In this section, we will create five methods that each accept the same set of data and perform the same filtering on the input data. However, the ordering in each PLINQ query will be handled differently. We are going to be working with the same `Person` class from the previous section. So, you can either work with the same project or create a new .NET console application project and add the `People` class from the previous example. Let's get started:

1. First, open the `Person` class and add a new `bool` property named `IsImportant`. We are going to use this to add a second data point for filtering in the PLINQ queries:

    ```
    internal class Person
    {
        public string FirstName { get; set; } = "";
        public string LastName { get; set; } = "";
        public int Age { get; set; }
        public bool IsImportant { get; set; }
    }
    ```

2. Next, add a new class to the project named `OrderSamples`.

3. Now it's time to start adding the queries. In this first query, we are not specifying `AsOrdered` or `AsUnordered`. By default, PLINQ should not be attempting to preserve the original order of the data. In each of these queries, we are returning each `Person` object with `Age` less than 18 and with `IsImportant` set to `true`:

    ```
    internal IEnumerable<Person>
        GetImportantChildrenNoOrder(List<Person> people)
    {
    ```

```
        return people.AsParallel()
            .Where(p => p.IsImportant && p.Age < 18);
    }
```

4. In the second example, we are explicitly adding `IsUnordered` to the query after `AsParallel`. The behavior should be the same as the first query, with PLINQ not concerning itself with the original order of the items:

```
    internal IEnumerable<Person>
        GetImportantChildrenUnordered(List<Person> people)
    {
        return people.AsParallel().AsUnordered()
            .Where(p => p.IsImportant && p.Age < 18);
    }
```

5. The third example breaks up the filters into two separate `Where` clauses; `IsSequential` is added after the first `Where` clause. How do you think this will impact the item sequence? We will find out when we run the program:

```
    internal IEnumerable<Person>
        GetImportantChildrenUnknownOrder(List<Person>
            people)
    {
        return people.AsParallel().Where(p =>
            p.IsImportant)
            .AsSequential().Where(p => p.Age < 18);
    }
```

6. In the fourth example, we are using `AsParallel().AsOrdered()` to signal to PLINQ that we want the original order of items to be preserved:

```
    internal IEnumerable<Person>
        GetImportantChildrenPreserveOrder(List<Person>
            people)
    {
        return people.AsParallel().AsOrdered()
            .Where(p => p.IsImportant && p.Age < 18);
    }
```

7. In the fifth and final example, we are adding a `Reverse` method after `AsOrdered`. This
    should preserve the original order of items in reverse:

```
internal IEnumerable<Person>
    GetImportantChildrenReverseOrder(List<Person>
        people)
{
    return people.AsParallel().AsOrdered().Reverse()
        .Where(p => p.IsImportant && p.Age < 18);
}
```

8. Next, open `Program.cs` and add two local functions. One will create a list of `Person`
    objects to pass to each method. The other will iterate over `List<Person>` to output each
    `FirstName` to the console:

```
static List<Person> GetYoungPeople()
{
    return new List<Person>
    {
        new Person { FirstName = "Bob", LastName =
            "Jones", Age = 23 },
        new Person { FirstName = "Sally", LastName =
            "Shen", Age = 2, IsImportant = true },
        new Person { FirstName = "Joe", LastName =
            "Smith", Age = 5, IsImportant = true },
        new Person { FirstName = "Lisa", LastName =
            "Samson", Age = 9, IsImportant = true },
        new Person { FirstName = "Norman", LastName =
            "Patterson", Age = 17 },
        new Person { FirstName = "Steve", LastName =
            "Gates", Age = 20 },
        new Person { FirstName = "Richard", LastName =
            "Ng", Age = 16, IsImportant = true }
    };
}
static void OutputListToConsole(List<Person> list)
{
    foreach (var item in list)
```

```
        {
            Console.WriteLine(item.FirstName);
        }
    }
}
```

9. Finally, we will add the code to call each method. The timestamp, including milliseconds, is being output to the console before each method call and again at the end. You can run the application multiple times to inspect the performance of each method call. Try running it on PCs with more or fewer cores and different-sized datasets to see how that impacts the output:

```
using LINQandPLINQsnippets;
var timeFmt = "hh:mm:ss.fff tt";
var orderExample = new OrderSamples();
Console.WriteLine($"Start time: {DateTime.Now.ToString
    (timeFmt)}. AsParallel children:");
OutputListToConsole(orderExample.GetImportantChildrenN
    oOrder(GetYoungPeople()).ToList());
Console.WriteLine($"Start time: {DateTime.Now
    .ToString(timeFmt)}. AsUnordered children:");
OutputListToConsole(orderExample.GetImportantChildrenU
    nordered(GetYoungPeople()).ToList());
Console.WriteLine($"Start time: {DateTime.Now
    .ToString(timeFmt)}. Sequential after Where
        children:");
OutputListToConsole(orderExample.GetImportantChildren
    UnknownOrder(GetYoungPeople()).ToList());
Console.WriteLine($"Start time: {DateTime.Now
    .ToString(timeFmt)}. AsOrdered children:");
OutputListToConsole(orderExample.GetImportantChildrenP
    reserveOrder(GetYoungPeople()).ToList());
Console.WriteLine($"Start time: {DateTime.Now
    .ToString(timeFmt)}. Reverse order children:");
OutputListToConsole(orderExample.GetImportantChildrenR
    everseOrder(GetYoungPeople()).ToList());
Console.WriteLine($"Finish time: {DateTime.Now
    .ToString(timeFmt)}");
Console.ReadLine();
```

10. Now, run the program and examine the output:

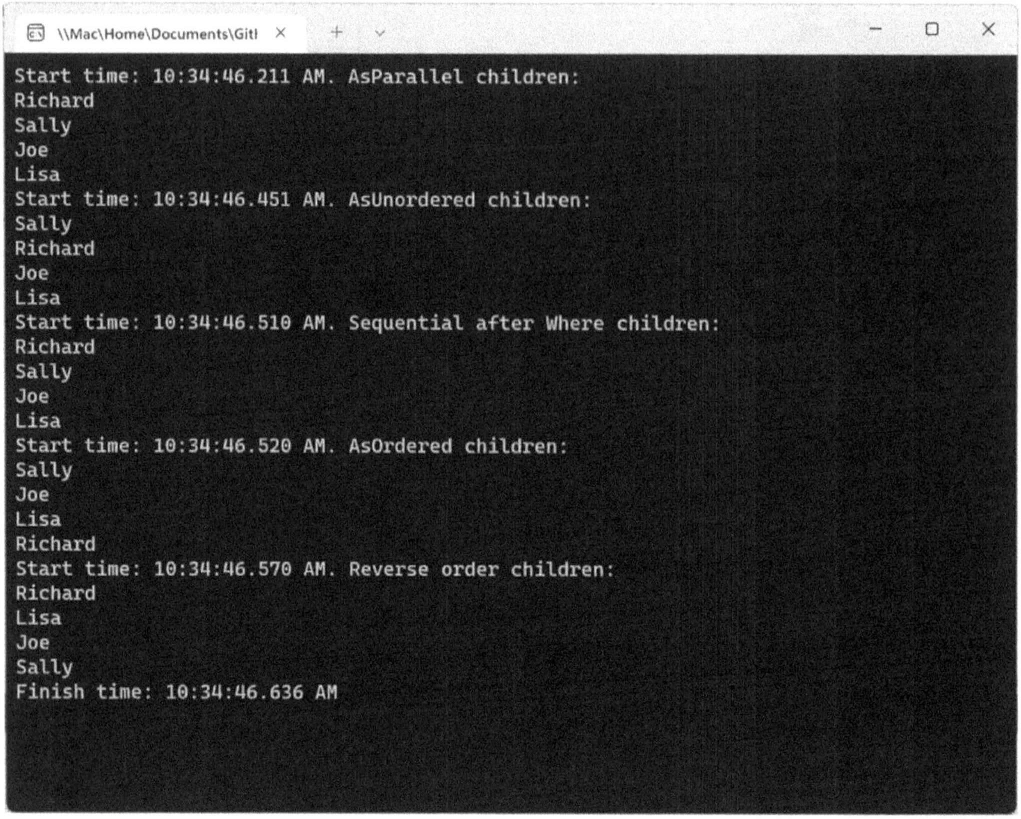

```
Start time: 10:34:46.211 AM. AsParallel children:
Richard
Sally
Joe
Lisa
Start time: 10:34:46.451 AM. AsUnordered children:
Sally
Richard
Joe
Lisa
Start time: 10:34:46.510 AM. Sequential after Where children:
Richard
Sally
Joe
Lisa
Start time: 10:34:46.520 AM. AsOrdered children:
Sally
Joe
Lisa
Richard
Start time: 10:34:46.570 AM. Reverse order children:
Richard
Lisa
Joe
Sally
Finish time: 10:34:46.636 AM
```

Figure 8.3 – Comparing the order of items from five PLINQ queries

You can see from the output that the order of items is only predictable in the last two examples where we have specified AsOrdered() and AsOrdered().Reverse(). The impact of the different PLINQ operations is difficult to measure on such a small dataset. If you run this several times, you are likely to see different results in the timing. Try adding larger datasets on your own to experiment with the performance.

Next, let's discuss merging segments and test the different options in a sample.

Using WithMergeOptions in PLINQ queries

When we discuss merging data in PLINQ, it is the merge that happens as each segment of an operation completes its actions, and the results are merged back into the result on the calling thread. Most of the time, you will not need to specify any merge options. For times when you may need to do so, it's important to understand the behavior of each of the options. Let's review each of the members of the `ParallelMergeOptions` enumeration.

ParallelMergeOptions.NotBuffered

Think of the `NotBuffered` option as streaming data. Each item is returned from the query immediately after it has finished processing. There are some PLINQ operations that cannot support this option and will ignore it. For instance, the `OrderBy` and `OrderByDescending` operations cannot return items until the sorting has completed on the merged data. These are always `FullyBuffered`. However, queries that use `AsOrdered` can use this option. Use this option if your application needs to consume items in a streaming manner.

ParallelMergeOptions.AutoBuffered

The `AutoBuffered` option returns sets of items as they are collected. The size of the item set and how frequently it is returned to clear the buffer are not configurable or known to your code. If you want to make your data available in this manner, this option may suit your needs. Once again, the `OrderBy` and `OrderByDescending` operations will not accept this option. This is the default for most PLINQ operations and is the fastest overall in most scenarios. The `AutoBuffered` option allows PLINQ the most flexibility to buffer items as necessary based on current system conditions.

ParallelMergeOptions.FullyBuffered

The `FullyBuffered` option will not make any results available until they have all been processed and buffered by the query. The option will take the longest to make the first item available, but many times, it is the fastest to provide the entire dataset.

ParallelMergeOptions.Default

There is also the `ParallelMergeOptions.Default` value, which will act the same as not calling `WithMergeOptions` at all. You should choose your merge option based on how the data needs to be consumed. If you have no strict requirements, it is usually best to not set merge options.

WithMergeOptions in action

Let's create examples of using the same `Person` query with each merge option and with no merge options set at all:

1. Start by adding a `MergeSamples` class to the console application project you previously created. First, add the following three methods to test the types of merges:

    ```
    internal IEnumerable<Person>
        GetImportantChildrenNoMergeSpecified(List<Person>
            people)
    {
    ```

```
        return people.AsParallel()
            .Where(p => p.IsImportant && p.Age < 18)
                .Take(3);
    }
    internal IEnumerable<Person> GetImportantChildren
        DefaultMerge(List<Person> people)
    {
        return people.AsParallel().WithMergeOptions
            (ParallelMergeOptions.Default)
                .Where(p => p.IsImportant && p.Age <
                    18).Take(3);
    }
    internal IEnumerable<Person> GetImportant
        ChildrenAutoBuffered(List<Person> people)
    {
        return people.AsParallel().WithMergeOptions
            (ParallelMergeOptions.AutoBuffered).Where(p =>
                p.IsImportant && p.Age < 18).Take(3);
    }
```

2. Next, add the following two methods to the `MergeSamples` class:

```
    internal IEnumerable<Person> GetImportant
        ChildrenNotBuffered(List<Person> people)
    {
        return people.AsParallel().WithMergeOptions
            (ParallelMergeOptions.NotBuffered)
                .Where(p => p.IsImportant && p.Age <
                    18).Take(3);
    }
    internal IEnumerable<Person> GetImportantChildren
        FullyBuffered(List<Person> people)
    {
        return people.AsParallel().WithMergeOptions
            (ParallelMergeOptions.FullyBuffered).Where(p =>
                p.IsImportant && p.Age < 18).Take(3);
    }
```

Each of the methods in the last two steps performs a PLINQ query that filters for IsImportant equal to true and Age less than 18. It then performs a Take(3) operation to return only the first three items from the query.

3. Add code to Program.cs to call each method and output the timestamp before each call, as well as a final timestamp at the end. This is the same process we used when calling the methods to test ordering in the previous section:

```
using LINQandPLINQsnippets;
var timeFmt = "hh:mm:ss.fff tt";
var mergeExample = new MergeSamples();
Console.WriteLine($"Start time: {DateTime.Now.ToString
    (timeFmt)}. NoMerge children:");
OutputListToConsole(mergeExample.GetImportantChildrenN
    oMergeSpecified(GetYoungPeople()).ToList());
Console.WriteLine($"Start time:
    {DateTime.Now.ToString(timeFmt)}. DefaultMerge
        children:");
OutputListToConsole(mergeExample.GetImportantChildren
    DefaultMerge(GetYoungPeople()).ToList());
Console.WriteLine($"Start time: {DateTime.Now.ToString
    (timeFmt)}. AutoBuffered children:");
OutputListToConsole(mergeExample.GetImportantChildren
    AutoBuffered(GetYoungPeople()).ToList());
Console.WriteLine($"Start time:
    {DateTime.Now.ToString(timeFmt)}. NotBuffered
        children:");
OutputListToConsole(mergeExample.GetImportantChildren
    NotBuffered(GetYoungPeople()).ToList());
Console.WriteLine($"Start time:
    {DateTime.Now.ToString(timeFmt)}. FullyBuffered
        children:");
OutputListToConsole(mergeExample.GetImportantChildren
    FullyBuffered(GetYoungPeople()).ToList());
Console.WriteLine($"Finish time: {
        DateTime.Now.ToString(timeFmt)}");
Console.ReadLine();
```

4.   Now, run the program and examine the output:

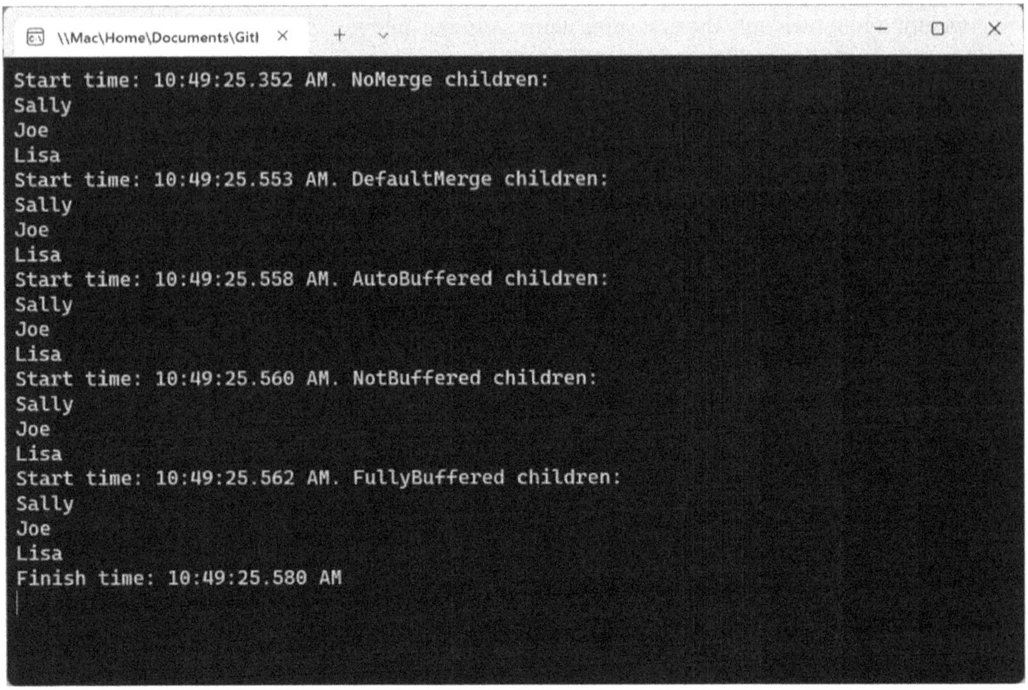

```
Start time: 10:49:25.352 AM. NoMerge children:
Sally
Joe
Lisa
Start time: 10:49:25.553 AM. DefaultMerge children:
Sally
Joe
Lisa
Start time: 10:49:25.558 AM. AutoBuffered children:
Sally
Joe
Lisa
Start time: 10:49:25.560 AM. NotBuffered children:
Sally
Joe
Lisa
Start time: 10:49:25.562 AM. FullyBuffered children:
Sally
Joe
Lisa
Finish time: 10:49:25.580 AM
```

Figure 8.4 – Reviewing the output of the PLINQ merge options methods

The first option with no merge option specified takes the longest to run, but often, the first time you run a PLINQ query, it will be slower than subsequent executions. The remaining queries are all very fast. You should test these queries on some large sets of data from your own databases and see how the timings differ for different PLINQ operators and different merge options. You can even take timings between the output of each item to see how quickly the first item is returned for NotBuffered versus FullyBuffered.

Before we review everything that we have learned in this chapter, let's discuss a few .NET objects and data structures that complement parallel programming and PLINQ queries.

Data structures for parallel programming in .NET

When working with parallel programming in .NET, and with PLINQ, you should take advantage of the data structures, types, and primitives that .NET provides. In this section, we will touch on concurrent collections and **synchronization primitives**.

Concurrent collections

Concurrent collections are useful when working with parallel programming. We will cover them in great detail in *Chapter 9*, but let's quickly discuss how we can leverage them when working with PLINQ queries.

If you are simply selecting and sorting data with PLINQ, it is not necessary to incur the overhead that is added with the collections in the `System.Collections.Concurrent` namespace. However, if you are calling a method with `ForAll` that modifies items in your source data, you should use one of these current collections, such as `BlockingCollection<T>`, `ConcurrentBag<T>`, or `ConcurrentDictionary<TKey, TValue>`. They can also guard against any simultaneous `Add` or `Remove` operations on the collections.

Synchronization primitives

If you are unable to introduce concurrent collections into your existing code base, another option to provide concurrency and performance is synchronization primitives. We covered many of these types in *Chapter 1*. These types in the `System.Threading` namespace, including `Barrier`, `CountdownEvent`, `SemaphoreSlim`, `SpinLock`, and `SpinWait`, provide the right balance of thread safety and performance. Other locking mechanisms, such as `lock` and `Mutex`, can be more expensive to implement, causing a greater performance impact.

If we want to guard one of our PLINQ queries that uses `ForAll` with `SpinLock`, we can simply wrap the method in a `try/finally` block and use the `Enter` and `Exit` calls on `SpinLock`. Take this example where we were checking where a person had an age greater than `120`. Let's imagine that the code also modifies the age:

```
private SpinLock _spinLock = new SpinLock();
internal void ProcessAdultsWhoVoteWithPlinq2(List<Person>
    people)
{
    var adults = people.AsParallel().Where(p => p.Age > 17);
    adults.ForAll(ProcessVoterActions2);
}
private void ProcessVoterActions2(Person adult)
{
    var hasLock = false;
    if (adult.Age > 120)
    {
        try
```

```
        {
            _spinLock.Enter(hasLock);
            adult.Age = 120;
        }
        finally
        {
            if (hasLock) _spinLock.Exit();
        }
    }
}
```

To read more about synchronization primitives, check out this section in Microsoft Docs: `https://docs.microsoft.com/dotnet/standard/threading/overview-of-synchronization-primitives`.

Now, let's wrap up by reviewing what we have learned in this chapter on parallel programming and PLINQ.

Summary

In this chapter, we learned about the power of PLINQ to introduce parallel processing to our LINQ queries. We started by looking at how PLINQ differs from standard LINQ queries. Next, we explored how to introduce PLINQ into existing code by converting some standard LINQ queries. It is important to understand how PLINQ is impacting the performance of your applications, and we examined some timings in our sample applications. (Later, in *Chapter 10*, we will discuss some tools to test your application performance while testing it locally.) We covered some optimizations you can make to your queries with merge options and data ordering. Finally, we wrapped up by touching on some other .NET data structures and types to help provide type safety and performance to your applications.

In the next chapter, we will explore each of the concurrent collections in the `System.Collections.Concurrent` namespace in depth. The concurrent collections are key to ensuring that your parallel and concurrent code maintains type safety when operating on shared data.

Questions

1. Which PLINQ method signals that the query should start processing in parallel?

2. Which PLINQ method signals that the query should not process in parallel any longer?

3. Which method tells PLINQ to preserve the original order of the source data?

4. Which PLINQ method will execute a delegate in parallel on each item in the query?

5. What performance impact does `AsOrdered()` have on a PLINQ query?

6. Which PLINQ operations cannot be used with `ParallelMergeOptions.NotBuffered`?

7. Is PLINQ always faster than an equivalent LINQ query?

8. Which PLINQ merge option would you select if you want results to stream back from the query as they become available?

9

Working with Concurrent Collections in .NET

This chapter will dive deeper into some of the **concurrent collections** in the System.Collections. Concurrent namespace. These specialized collections help to preserve data integrity when using concurrency and parallelism in your C# code. Each section of this chapter will provide practical examples of how to use a specific concurrent collection provided by .NET.

We have seen some basic use of parallel data structures in .NET. We have already covered the basics of each of the concurrent collections in the *Introduction to concurrency* section of *Chapter 2*. So, we will quickly jump into the examples of their use in this chapter and learn more about their application and inner workings.

In this chapter, we will do the following:

- Using BlockingCollection
- Using ConcurrentBag
- Using ConcurrentDictionary
- Using ConcurrentQueue
- Using ConcurrentStack

By the end of this chapter, you will have a deeper understanding of how these collections protect your shared data from being mishandled while multithreading.

Technical requirements

To follow along with the examples in this chapter, the following software is recommended for Windows developers:

- Visual Studio 2022 version 17.0 or later.

- .NET 6.

- To complete any WinForms or WPF samples, you will need to install the .NET desktop development workload for Visual Studio. These projects will run only on Windows.

While these are recommended, if you have .NET 6 installed, you can use your preferred editor. For example, Visual Studio 2022 for Mac on macOS 10.13 or later, JetBrains Rider, or Visual Studio Code will work just as well.

All the code examples for this chapter can be found on GitHub at `https://github.com/ PacktPublishing/Parallel-Programming-and-Concurrency-with-C- sharp-10-and-.NET-6/tree/main/chapter09`.

Let's get started by learning more about `BlockingCollection<T>` and walk through a sample project that leverages the collection.

Using BlockingCollection

`BlockingCollection<T>` is one of the most useful concurrent collections. As we saw in *Chapter 7*, `BlockingCollection<T>` was created to be an implementation of the **producer/ consumer pattern** for .NET. Let's review some of the specifics of this collection before creating a different kind of sample project.

BlockingCollection details

One of the major draws of `BlockingCollection<T>` for developers working with parallel code implementations is that it can be swapped to replace `List<T>` without too many additional modifications. You can use the `Add()` method for both. The difference with `BlockingCollection<T>` is that calling `Add()` to add an item will block the current thread if another read or write operation is in process. If you want to specify a timeout period on the operation, you can use `TryAdd()`. The `TryAdd()` method optionally supports both timeouts and cancellation tokens.

Removing items from `BlockingCollection<T>` with `Take()` has an equivalent `TryTake()`, which allows timed operations and cancellation. The `Take()` and `TryTake()` methods will take and remove the first remaining item that was added to the collection. This is because the default underlying collection type within `BlockingCollection<T>` is `ConcurrentQueue<T>`. Alternatively, you can specify that the collection uses `ConcurrentStack<T>`, `ConcurrentBag<T>`, or any collection that implements the `IProducerConsumerCollection<T>` interface. Here's an example of `BlockingCollection<T>` being initialized to use `ConcurrentStack<T>`, with its capacity limited to `100` items:

```
var itemCollection = new BlockingCollection<string>(new
    ConcurrentStack<string>(), 100);
```

If your application needs to iterate over the items in `BlockingCollection<T>`, the `GetConsumingEnumerable()` method can be used in a `for` or `foreach` loop. However, keep in mind that this iteration over the collection is also removing items, and it will complete the collection if the enumeration continues until the collection is empty. This is the *consuming* part of the `GetConsumingEnumerable()` method name.

If you need to use multiple `BlockingCollection<T>` classes of the same type, you can add to or take from them as one by adding them to an array. An array of `BlockingCollection<T>` makes the `TryAddToAny()` and `TryTakeFromAny()` methods available. These methods will succeed if any of the collections in the array are in the proper state to accept or provide objects to the calling code. Microsoft Docs has an example of how to use an array of `BlockingCollection<T>` in a pipeline: `https://docs.microsoft.com/dotnet/standard/collections/thread-safe/how-to-use-arrays-of-blockingcollections`.

Now that we have covered the details needed to understand `BlockingCollection<T>`, let's dive into a sample project.

Using BlockingCollection with Parallel.ForEach and PLINQ

We already covered an example that implements the producer/consumer pattern in *Chapter 7*, so let's try something a little different in this section. We are going to create a WPF application that loads the contents of a book from a 1.5 MB text file and searches for words that start with a particular letter:

> **Note**
>
> This sample uses a .NET Standard NuGet package created from a Microsoft extension sample that was originally built on .NET Framework 4.0. The extension is called `ParallelExtensionsExtras`, and the original source is available on GitHub: `https://github.com/dotnet/samples/tree/main/csharp/parallel/ParallelExtensionsExtras`. The extension method that we will be using from the package makes `Parallel.ForEach` operations and PLINQ queries run more efficiently with concurrent collections. To read more about the extensions, you can check out this post on the *.NET Parallel Programming* blog: `https://devblogs.microsoft.com/pfxteam/parallelextensionsextras-tour-4-blockingcollectionextensions/`.

1. Start by creating a new WPF application in Visual Studio. Name the project `ParallelExtras.BlockingCollection`.

2. On the NuGet Package Manager page, search for and add the latest stable version of the **ParallelExtensionsExtras.NetFxStandard** package to your project:

Figure 9.1 – The ParallelExtensionsExtras.NetFxStandard NuGet package

3. We are going to read text from the book *Ulysses* by James Joyce. This book is public domain in the United States and most countries around the world. It can be downloaded in UTF-8 plain text format from **Project Gutenberg**: `https://www.gutenberg.org/ebooks/4300`. Download a copy, name the file `ulysses.txt`, and place it in the main folder with your other project files.

4. In Visual Studio, right-click `ulysses.txt` and select **Properties**. In the **Properties** window, update the **Copy to Output Directory** property to **Copy if newer**.

5. Open **MainWindow.xaml** and add `Grid.RowDefinitions` and `Grid.Columndefinitions` to the `Grid` control, as follows:

```
<Grid.RowDefinitions>
    <RowDefinition Height="Auto"/>
    <RowDefinition Height="*"/>
</Grid.RowDefinitions>
<Grid.ColumnDefinitions>
    <ColumnDefinition/>
    <ColumnDefinition/>
</Grid.ColumnDefinitions>
```

6. Add `ComboBox` and `Button` inside the `Grid` definition following the `Grid.ColumnDefinitions` element. These controls will be in the first row of `Grid`:

```
<ComboBox x:Name="LettersComboBox"
            Grid.Row="0" Grid.Column="0"
            Margin="4">
    <ComboBoxItem Content="A"/>
    <ComboBoxItem Content="D"/>
    <ComboBoxItem Content="F"/>
    <ComboBoxItem Content="G"/>
    <ComboBoxItem Content="M"/>
    <ComboBoxItem Content="O"/>
```

```
        <ComboBoxItem Content="A"/>
        <ComboBoxItem Content="T"/>
        <ComboBoxItem Content="W"/>
    </ComboBox>
    <Button Grid.Row="0" Grid.Column="1"
            Margin="4" Content="Load Words"
            Click="Button_Click"/>
```

ComboBox will contain nine different letters from which to choose. You can add as many of these as you like. Button contains a Click event handler that we will add to MainWindow.xaml.cs soon.

7. Finally, add ListView named WordsListView to the second row of Grid. It will span both of the columns:

```
    <ListView x:Name="WordsListView" Margin="4"
              Grid.Row="1" Grid.ColumnSpan="2"/>
```

8. Now, open MainWIndow.xaml.cs. The first thing we will do here is to create a method named LoadBookLinesFromFile(), which reads each line of text from ulysses. txt into BlockingCollection<string>. There is only a single thread reading from the file, so using the Add() method instead of TryAdd() is best:

```
    private async Task<BlockingCollection<string>>
        LoadBookLinesFromFile()
    {
        var lines = new BlockingCollection<string>();
        using var reader = File.OpenText(Path.Combine(
            Path.GetDirectoryName(Assembly
                .GetExecutingAssembly().Location),
                    "ulysses.txt"));
        string line;
        while ((line = await reader.ReadLineAsync()) !=
            null)
        {
            lines.Add(line);
        }
        lines.CompleteAdding();
        return lines;
    }
```

> **Note**
>
> Remember, it is important to call `lines.CompleteAdding()` before the end of the
> method. Otherwise, subsequent queries of this collection will hang and continue waiting for
> more items to be added to the stream.

9. Now, create a method named `GetWords()` that takes the lines from the text file and uses
    a **regular expression** to parse each line into individual words. These words will all be added
    to a new `BlockingCollection<string>`. In this method, we're parsing multiple
    lines simultaneously with a `Parallel.ForEach` loop. The **ParallelExtentionsExtras.
    NetFxStandard** package provides the `GetConsumingPartitioner()` extension method,
    which tells the `Parallel.ForEach` loop that `BlockingCollection` will be doing its
    own blocking, so the loop does not need to do any. This makes the whole process more efficient:

```
private BlockingCollection<string>
    GetWords(BlockingCollection<string> lines)
{
    var words = new BlockingCollection<string>();
    Parallel.ForEach(lines.GetConsumingPartitioner(),
        (line) =>
    {
        var matches = Regex.Matches(line,
            @"\b[\w']*\b");
        foreach (var m in matches.Cast<Match>())
        {
            if (!string.IsNullOrEmpty(m.Value))
            {
                words.TryAdd(TrimSuffix(m.Value,
                    '\''));
            }
        }
    });
    words.CompleteAdding();
    return words;
}
private string TrimSuffix(string word, char
    charToTrim)
{
    int charLocation = word.IndexOf(charToTrim);
```

```
        if (charLocation != -1)
        {
            word = word[..charLocation];
        }
        return word;
}
```

The `TrimSuffix()` method will remove specific characters from the end of a word; in this case, we're passing the apostrophe character to be removed.

> **Note**
>
> If you are unfamiliar with regular expressions, you can read about how to use them with .NET on Microsoft Docs: `https://docs.microsoft.com/dotnet/standard/base-types/regular-expressions`. They are an extremely efficient way to parse text.

10. Next, create a method named `GetWordsByLetter()` to call the other methods we just created. Once `BlockingCollection<string>` containing all the words from the book has been fetched, this method will use PLINQ and `GetConsumingPartitioner()` to find all words that start with the uppercase or lowercase versions of the selected letter:

```
private async Task<List<string>> GetWordsByLetter(char
    letter)
{
    BlockingCollection<string> lines = await
        LoadBookLinesFromFile();
    BlockingCollection<string> words =
        GetWords(lines);
    // 275,506 words in total
    return words.GetConsumingPartitioner()
        .AsParallel()
        .Where(w => w.StartsWith(letter) ||
            w.StartsWith(char.ToLower(letter)))
        .ToList();
}
```

11. Finally, we'll add the `Button_Click` event to kick off the loading, parsing, and querying of the book's text. Don't forget to mark the event handler as `async`:

```
private async void Button_Click(object sender,
    RoutedEventArgs e)
```

```
    {
        if (LettersComboBox.SelectedIndex < 0)
        {
            MessageBox.Show("Please select a letter.");
            return;
        }
        WordsListView.ItemsSource = await
            GetWordsByLetter(
            char.Parse(GetComboBoxValue(LettersComboBox
                .SelectedValue)));
    }
    private string GetComboBoxValue(object item)
    {
        var comboxItem = item as ComboBoxItem;
        return comboxItem.Content.ToString();
    }
```

The GetComboBoxValue() helper method will take the object from
LettersComboBox.SelectedValue and find string with the selected letter
within.

12. The following using declarations are required in MainWindow.xaml.cs to compile
and run the project:

```
using System.Collections.Concurrent;
using System.Collections.Generic;
using System.IO;
using System.Linq;
using System.Reflection;
using System.Text.RegularExpressions;
using System.Threading.Tasks;
using System.Windows;
using System.Windows.Controls;
```

13. Now, run the project, select a letter, and click **Load Words**:

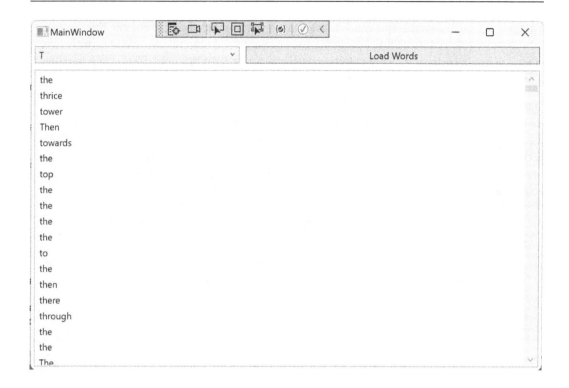

Figure 9.2 – Displaying words that begin with T from ulysses.txt

The whole process runs very quickly considering the book contains over 275,000 total words. Try adding some sorting to the PLINQ query and see how the performance is impacted.

Let's continue by learning about `ConcurrentBag<T>`.

Using ConcurrentBag

The `ConcurrentBag<T>` is an unordered collection of objects that can be safely added, peeked at, or removed concurrently. Keep in mind that, as with all of the concurrent collections, the methods exposed by `ConcurrentBag<T>` are thread-safe, but any extension methods are not guaranteed to be safe. Always implement your own synchronization when leveraging them. To review a list of safe methods, you can review this Microsoft Docs page: `https://docs.microsoft.com/dotnet/api/system.collections.concurrent.concurrentbag-1#methods`.

We are going to create a sample application that simulates working with a pool of objects. This scenario can be useful if you have some processing that leverages a stateful object that is memory-intensive. You want to minimize the number of objects created but cannot reuse one until the previous iteration has finished using it and returned it to the pool.

In our example, we will use a mocked-up PDF processing class that is assumed to be memory-intensive. In reality, document-processing libraries can be quite heavy, and they often rely on document states in each instance. The console application will iterate in parallel 15 times to create these fake PDF objects and append some text to each of them. Each time through the loop, we will output the text contents and the current count of PDF processors in the pool. If the current count remains low, then the application is working as intended:

1. Start by creating a new .NET console application in Visual Studio named `ConcurrentBag.PdfProcessor`.

2. Add a new class to represent the mocked-up PDF data. Name the class `ImposterPdfData`:

```
public class ImposterPdfData
{
    private string _plainText;
    private byte[] _data;
    public ImposterPdfData(string plainText)
    {
        _plainText = plainText;
        _data = System.Text.Encoding.ASCII.GetBytes
            (plainText);
    }
    public string PlainText => _plainText;
    public byte[] PdfData => _data;
}
```

We are storing the plain text and an ASCII-encoded version of the text that we will pretend is PDF format. This avoids implementing any third-party libraries in our sample application. If you have any PDF libraries with which you are familiar, you are welcome to adapt this sample to use them.

3. Next, add a new class named `PdfParser`. This class will be the one that is taken from and returned to `ConcurrentBag<PdfParser>`. We will create the host for that collection in an upcoming step:

```
public class PdfParser
{
    private ImposterPdfData? _pdf;
```

```
public void SetPdf(ImposterPdfData pdf) =>
    _pdf = pdf;
public ImposterPdfData? GetPdf() => _pdf;
public string GetPdfAsString()
{
    if (_pdf != null)
        return _pdf.PlainText;
    else
        return "";
}
public byte[] GetPdfBytes()
{
    if (_pdf != null)
        return _pdf.PdfData;
    else
        return new byte[0];
}
}
```

This stateful class holds an instance of an `ImposterPdfData` object and can return the data as a string or the ASCII-encoded byte array.

4. Add a method to `PdfParser` named `AppendString`. This method will add some additional text to `ImposterPdfData` on a new line:

```
public void AppendString(string data)
{
    string newData;
    if (_pdf == null)
    {
        newData = data;
    }
    else
    {
        newData = _pdf.PlainText + Environment.NewLine
            + data;
    }
```

```
        _pdf = new ImposterPdfData(newData);
    }
```

5. Now, add a class named `PdfWorkerPool`:

```
public class PdfWorkerPool
{
    private ConcurrentBag<PdfParser> _workerPool =
        new();
    public PdfWorkerPool()
    {
        // Add initial worker
        _workerPool.Add(new PdfParser());
    }
    public PdfParser Get() => _workerPool.TryTake(out
        var parser) ? parser : new PdfParser();
    public void Return(PdfParser parser) =>
        _workerPool.Add(parser);
    public int WorkerCount => _workerPool.Count();
}
```

Be sure to also add a `using System.Collections.Concurrent;` statement to `PdfWorkerPool.cs`. The pool stores `ConcurrentBag<PdfParser>` named `_workerPool`. When `PdfWorkerPool` is initialized, it adds a new instance to `_workerPool`. The `Get` method will return an existing instance from the pool with `TryTake` if one exists. If the pool is empty, a new instance is created and returned to the caller. The `Return` method adds `PdfParser` back to the pool when the consumer is finished. We will use the `WorkerCount` property to track the number of objects in the pool at any time.

6. Finally, replace the contents of Program.cs with the following code:

```
using ConcurrentBag.PdfProcessor;
Console.WriteLine("Hello, ConcurrentBag!");
var pool = new PdfWorkerPool();
Parallel.For(0, 15, async (i) =>
{
    var parser = pool.Get();
    var data = new ImposterPdfData($"Data index: {i}");
    try
    {
        parser.SetPdf(data);
        parser.AppendString(DateTime.UtcNow
            .ToShortDateString());
        Console.WriteLine($"
            {parser.GetPdfAsString()}");
        Console.WriteLine($"Parser count:
            {pool.WorkerCount}");
        await Task.Delay(100);
    }
    finally
    {
        pool.Return(parser);
        await Task.Delay(250);
    }
});
Console.WriteLine("Press the Enter key to exit.");
Console.ReadLine();
```

After creating a new PdfWorkerPool, we're using a Parallel.For loop to iterate 15 times. Each time through the loop, we get PdfParser, set the text, append DateTime.UtcNow, and write the contents to the console, along with the current count of parsers in the pool.

7. Run the application and examine the output:

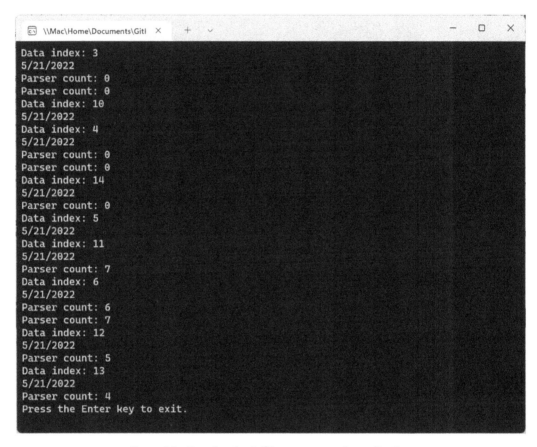

Figure 9.3 – Running the PdfProcessor console application

In my case, the parser count got to a maximum number of seven. If you tweak the `Task.Delay` intervals or remove them entirely, you are likely to see the count never exceed one. This kind of pool can be configured to be very efficient.

This application is an example where we do not care which instance of the collection is returned, so `ConcurrentBag<T>` is a perfect choice. In the next section, we will create a drug lookup example using `ConcurrentDictionary<TKey, TValue>`.

Using ConcurrentDictionary

In this section, we will create a WinForms application to load United States **Food and Drug Administration** (**FDA**) drug data concurrently from two files. Once loaded to `ConcurrentDictionary`, we can perform fast lookups with a **National Drug Code** (**NDC**) value to fetch the name. The FDA drug data is freely available to download in several formats from the NDC directory: `https://www.fda.gov/drugs/drug-approvals-and-databases/national-drug-code-`

`directory`. We will be working with tab-delimited text files. I have downloaded the `product.txt` file and moved about half of the records to a `product2.txt` file, duplicating the header row in the second file. You can get these files in the GitHub repository for the chapter at `https://github.com/PacktPublishing/Parallel-Programming-and-Concurrency-with-C-sharp-10-and-.NET-6/tree/main/chapter09/FdaNdcDrugLookup`:

1. Start by creating a new WinForms project in Visual Studio targeting .NET 6. Name the project `FdaNdcDrugLookup`.

2. Open the WinForm designer for `Form1.cs`. Lay out two `TextBox` controls, two `Button` controls, and `Label`:

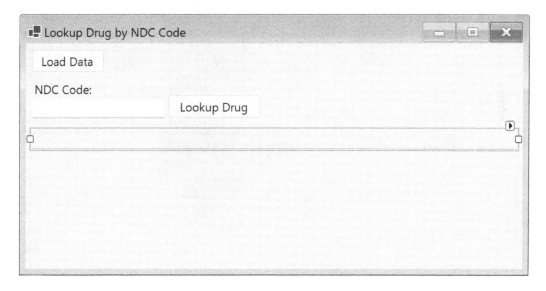

Figure 9.4 – The layout of Form1.cs

The **Load Data** button will have the following properties set: **Name** – `btnLoad` and **Text** – **Load Data**. The **NDC Code** text field will be named `txtNdc`. The **Lookup Drug** button will have these properties set: **Name** – `btnLookup`, **Text** – **Lookup Drug**, and **Enabled** – **False**. Finally, the **Drug Name** text field will have these properties set: **Name** – `txtDrugName` and **ReadOnly** – **True**.

3. Next, add the `product.txt` and `product2.txt` files to your project by right-clicking the project in **Solution Explorer** and choosing **Add | Existing Item**.

4. In the **Properties** panel, change **Copy to Output Directory** to **Copy if newer** for both of the text files we just added.

5. Add a new class to the project named `Drug` and add the following implementation:

```
public class Drug
{
    public string? Id { get; set; }
    public string? Ndc { get; set; }
    public string? TypeName { get; set; }
    public string? ProprietaryName { get; set; }
    public string? NonProprietaryName { get; set; }
    public string? DosageForm { get; set; }
    public string? Route { get; set; }
    public string? SubstanceName { get; set; }
}
```

This will contain the data for each record loaded from the NDC drug files.

6. Next, add a class to the project named `DrugService` and begin with the following implementation. To start, we only have `private ConcurrentDictionary<string, Drug>`. We will add a method to load the data in the next step:

```
using System.Collections.Concurrent;
using System.Data;
using System.Reflection;
namespace FdaNdcDrugLookup
{
    public class DrugService
    {
        private ConcurrentDictionary<string, Drug>
            _drugData = new();
    }
}
```

7. Next, add a public method to `DrugService` named `LoadData`:

```
public void LoadData(string fileName)
{
    using DataTable dt = new();
    using StreamReader sr = new(Path.Combine(
        Path.GetDirectoryName(Assembly
            .GetExecutingAssembly().Location),
```

```
                    fileName));
        var del = new char[] { '\t' };
        string[] colheaders = sr.ReadLine().Split(del);
        foreach (string header in colheaders)
        {
            dt.Columns.Add(header); // add headers
        }
        while (sr.Peek() > 0)
        {

            DataRow dr = dt.NewRow(); // add rows
            dr.ItemArray = sr.ReadLine().Split(del);
            dt.Rows.Add(dr);
        }
        foreach (DataRow row in dt.Rows)
        {

            Drug drug = new(); // map to Drug object
            foreach (DataColumn column in dt.Columns)
            {
                switch (column.ColumnName)
                {
                    case "PRODUCTID":
                        drug.Id = row[column].ToString();
                        break;
                    case "PRODUCTNDC":
                        drug.Ndc = row[column].ToString();
                        break;
...
// REMAINING CASE STATEMENTS IN GITHUB
                }
            }
            _drugData.TryAdd(drug.Ndc, drug);
        }
    }
}
```

> **Note**
>
> The `switch` statement in the previous snippet is truncated. To get the full code listing, visit the sample in the chapter's GitHub repository: `https://github.com/PacktPublishing/Parallel-Programming-and-Concurrency-with-C-sharp-10-and-.NET-6/tree/main/chapter09/FdaNdcDrugLookup`.

In this method, we are loading data from the provided `fileName` to `StreamReader`, adding the column headers to `DataTable`, populating its rows from the file, and then iterating over the rows and columns of `DataTable` to create `Drug` objects. Each `Drug` object is added to `ConcurrentDictionary` with a call to `TryAdd`, using the `Ndc` property as the key.

8. Now, add a `GetDrugByNdc` method to `DrugService` to complete the class. This method will return `Drug` for the provided `ndcCode`, if found:

    ```
    public Drug GetDrugByNdc(string ndcCode)
    {
        bool result = _drugData.TryGetValue(ndcCode, out
            var drug);
        if (result && drug != null)
            return drug;
        else
            return new Drug();
    }
    ```

9. Open the code for `Form1.cs` and add a private variable for the `DrugService`:

    ```
    private DrugService _drugService = new();
    ```

10. Open the designer for `Form1.cs` and double-click the **Load Data** button to create the `btnLoad_Click` event handler. Add the following implementation. Note that we made the `async` event handler to allow us to use the `await` keyword:

    ```
    private async void btnLoad_Click(object sender,
        EventArgs e)
    {
        var t1 = Task.Run(() => _drugService.LoadData
            ("product.txt"));
        var t2 = Task.Run(() => _drugService.LoadData
            ("product2.txt"));
        await Task.WhenAll(t1, t2);
    ```

```
    btnLookup.Enabled = true;
    btnLoad.Enabled = false;
}
```

To load the two text files, we are creating two tasks to run in parallel before using `Task.WhenAll` to await them both. Then, we can safely enable the `btnLookup` button and disable the `btnLoad` button to prevent a second load.

11. Next, switch back to the designer view for `Form1.cs` and double-click the **Lookup Drug** button. This will create the `btnLookup_Click` event handler. Add the following implementation to that handler to find a drug name based on the NDC code entered in the UI:

```
private void btnLookup_Click(object sender,
    EventArgs e)
{
    if (!string.IsNullOrWhiteSpace(txtNdc.Text))
    {
        var drug = _drugService.GetDrugByNdc
            (txtNdc.Text);
        txtDrugName.Text = drug.ProprietaryName;
    }
}
```

12. Now, run the application and click the **Load Data** button. After the load process has completed and the **Lookup Drug** button is enabled, enter the `70518-1120` NDC code. Click **Lookup Drug**:

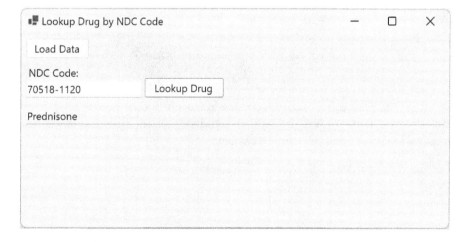

Figure 9.5 – Looking up the drug Prednisone by its NDC code

13. Try some other NDC codes and see how quickly each record loads. Here are a few random NDC codes taken from each file. If they all succeed, you know that both files loaded successfully in parallel: **0002-0800**, **0002-4112**, **43063-825**, and **51662-1544**.

That's it! You now have your own quick-and-dirty drug lookup application. Try replacing the drug name `TextBox` with `DataGrid` on your own to display an entire `Drug` record.

In the next section, we will work with customer orders in `ConcurrentQueue<T>`.

Using ConcurrentQueue

In this section, we will create a sample project that is a simplified version of a realistic scenario. We are going to create an order queuing system using `ConcurrentQueue<T>`. This application will be a console application that enqueues orders for two customers in parallel. We will create five orders for each customer, and to mix up the order of the queue, each customer queuing process will use a different `Task.Delay` between calls to `Enqueue`. The final output should show a mix of orders dequeued for the first customer and the second customer. Remember that `ConcurrentQueue<T>` employs **first in, first out** (**FIFO**) logic:

1. Let's start by opening Visual Studio and creating a .NET console application named `ConcurrentOrderQueue`.

2. Add a new class to the project named `Order`:

```
public class Order
{
    public int Id { get; set; }
    public string? ItemName { get; set; }
    public int ItemQty { get; set; }
    public int CustomerId { get; set; }
    public decimal OrderTotal { get; set; }
}
```

3. Now, create a new class named `OrderService` containing a private `ConcurrentQueue<Order>` named `_orderQueue`. This class is where we will enqueue and dequeue orders for our two customers:

```
using System.Collections.Concurrent;
namespace ConcurrentOrderQueue
{
    public class OrderService
    {
```

```
        private ConcurrentQueue<Order> _orderQueue =
            new();
    }
}
```

4. Let's start with the implementation of DequeueOrders. In this method, we will use a while loop to call TryDequeue until the collection is empty, adding each order to List<Order> to be returned to the caller:

```
public List<Order> DequeueOrders()
{
    List<Order> orders = new();
    while (_orderQueue.TryDequeue(out var order))
    {
        orders.Add(order);
    }
    return orders;
}
```

5. Now, we will create public and private EnqueueOrders methods. The public parameterless method will call the private method twice, once for each customerId. The two calls will be made in parallel, followed by a Task.WhenAll call to await them:

```
public async Task EnqueueOrders()
{
    var t1 = EnqueueOrders(1);
    var t2 = EnqueueOrders(2);
    await Task.WhenAll(t1, t2);
}
private async Task EnqueueOrders(int customerId)
{
    for (int i = 1; i < 6; i++)
    {
        var order = new Order
        {
            Id = i * customerId,
            CustomerId = customerId,
            ItemName = "Widget for customer " +
                customerId,
```

```
                        ItemQty = 20 - (i * customerId)
                };
                order.OrderTotal = order.ItemQty * 5;
                _orderQueue.Enqueue(order);
                await Task.Delay(100 * customerId);
            }
        }
```

The private `EnqueueOrders` method iterates five times to create and `Enqueue` orders for the given `customerId`. This is also used to vary `ItemName`, `ItemQty`, and the duration of `Task.Delay`.

6. Finally, open `Program.cs` and add the following code to enqueue and dequeue the orders, and output the resulting list to the console:

```
using ConcurrentOrderQueue;
Console.WriteLine("Hello, World!");
var service = new OrderService();
await service.EnqueueOrders();
var orders = service.DequeueOrders();
foreach(var order in orders)
{
    Console.WriteLine(order.ItemName);
}
```

7. Run the program and view the list of orders in the output. How does yours match up?

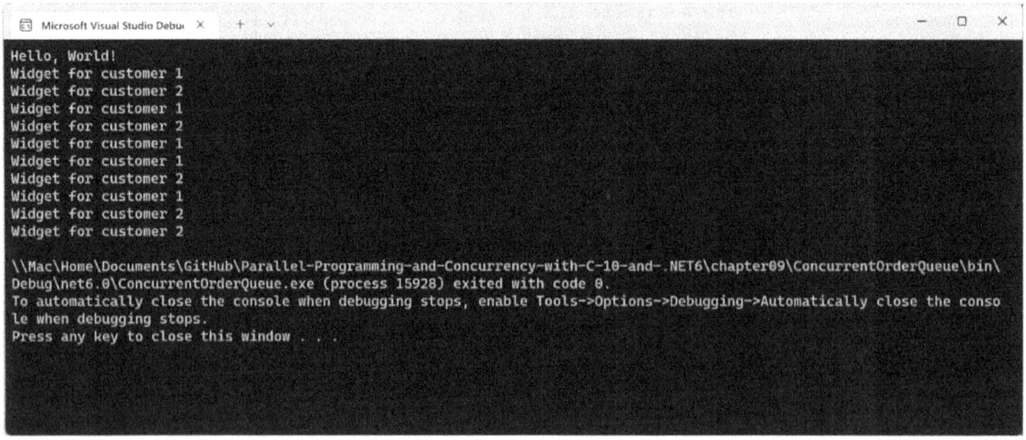

Figure 9.6 – Viewing the output of the order queue

Try varying the delay factor or changing `customerId` for one or both customers in the `EnqueueOrders` method to see how the order of the output changes.

Next, in the final section of the chapter, we will perform a quick experiment with the `ConcurrentStack<T>` collection.

Using ConcurrentStack

In this section, we are going to experiment with `BlockingCollection<T>` and `ConcurrentStack<T>`. In the first example in this chapter, we used `BlockingCollection<T>` to read the words that started with a specific letter from the book *Ulysses*. We are going to make a copy of that project and change the code that reads the lines of text to use `ConcurrentStack<T>` inside `BlockingCollection<T>`. This will make the lines output in reverse order because a stack uses **last in, first out** (**LIFO**) logic. Let's get started!

1. Make a copy of the **ParallelExtras.BlockingCollection** project from this chapter or modify the existing project if you prefer.

2. Open `MainWindow.xaml.cs` and modify the `LoadBookLinesFromFile` method to pass a new `ConcurrentStack<string>` to the constructor of `BlockingCollection<string>`:

    ```
    private async Task<BlockingCollection<string>>
        LoadBookLinesFromFile()
    {

        var lines = new BlockingCollection<string>(new
            ConcurrentStack<string>());

        ...

        return lines;
    }
    ```

    Note that the preceding method was truncated to emphasize the modified code. View the complete method on GitHub: `https://github.com/PacktPublishing/Parallel-Programming-and-Concurrency-with-C-sharp-10-and-.NET-6/tree/main/chapter09/ParallelExtras.ConcurrentStack`.

3. Now, when you run the application and search for the same letter as before (in our case, T), you will see a different set of words at the beginning of the list:

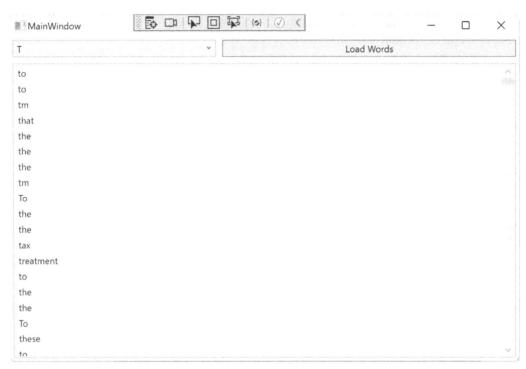

Figure 9.7 – Searching for words that start with T in Ulysses

If you scroll to the bottom of the list, you should see the words from the beginning of the book. Note that the list is not completely reversed because we didn't use `ConcurrentStack<string>` when parsing the words from each line. You can try this on your own as another experiment.

That concludes our tour of the .NET concurrent collections. Let's wrap up by summarizing what we have learned in this chapter.

Summary

In this chapter, we delved into five of the collections in the `System.Collections.Concurrent` namespace. We created five sample applications in the chapter to get some hands-on experience with each of the concurrent collection types available in .NET 6. Through a mix of WPF, WinForms, and .NET console application projects, we examined some real-world methods of leveraging these collections in your own applications.

In the next chapter, we will explore the rich set of tools provided by Visual Studio for multithreaded development and debugging. We will also discuss some techniques for analyzing and improving the performance of parallel .NET code.

Questions

1. Which concurrent collection can implement different types of collections under the covers?

2. What is the default internal collection type implemented by the collection in question 1?

3. Which collection type is frequently used as an implementation of the producer/consumer pattern?

4. Which concurrent collection contains key/value pairs?

5. Which method is used to add values to `ConcurrentQueue<T>`?

6. Which methods are used to add and get items in `ConcurrentDictionary`?

7. Are extension methods used with the concurrent collections thread-safe?

Part 3: Advanced Concurrency Concepts

In this part's chapters, you will gain the advanced skills you need to debug and test your parallel and concurrent code. You will also get some practical advice for safely canceling asynchronous work.

This part contains the following chapters:

- *Chapter 10, Debugging Multithreaded Applications with Visual Studio*
- *Chapter 11, Canceling Asynchronous Work*
- *Chapter 12, Unit Testing Async, Concurrent, and Parallel Code*

10

Debugging Multithreaded Applications with Visual Studio

Visual Studio 2022 is the latest version of Visual Studio on Mac and Windows. In this chapter, we are going to learn how to leverage the power of Visual Studio when debugging multithreaded .NET applications.

Visual Studio provides several extremely useful tools for developers who need to debug parallel and concurrent .NET applications. This chapter will explore the tools in detail through concrete examples.

In this chapter, we will cover the following topics:

- Introducing multithreaded debugging
- Debugging threads and processes
- Switching and flagging threads
- Debugging a parallel application

By the end of this chapter, you will have the tools and knowledge you need to debug threading issues in your parallel and concurrent C# code.

Technical requirements

To follow along with the examples in this chapter, the following software is recommended for Windows developers:

- Visual Studio 2022 version 17.2 or later

- .NET 6

- To complete any WinForms or WPF samples, you will need to install the .NET desktop development workload for Visual Studio. These projects will run only on Windows.

All the code examples for this chapter can be found on GitHub at `https://github.com/ PacktPublishing/Parallel-Programming-and-Concurrency-with-C- sharp-10-and-.NET-6/tree/main/chapter10`.

> **Note**
>
> The concepts and tools in this chapter only work with Visual Studio on Windows. If you are building .NET applications on a Mac, the **Rider** IDE from JetBrains provides several tools for multithreaded debugging – a **Threads** pane, a **Frames** view to view frames on a selected thread, and a **Parallel Stacks** pane. Visual Studio for Mac doesn't have this kind of support for debugging multithreaded applications yet. You can read more about JetBrains Rider's multithreaded debugging in their documentation: `https://www.jetbrains.com/ help/rider/Debugging_Multithreaded_Applications.html`. Debugging on a Mac is beyond the scope of this chapter.

Let's get started by learning some basics of multithreaded debugging with Visual Studio 2022.

Introducing multithreaded debugging

Debugging is a key component of every .NET developer's skillset. Nobody ever writes bug-free code and introducing multithreaded constructs to your project only increases the chances of introducing bugs. As .NET and C# have added more support for parallel programming and concurrency, Visual Studio has added more debugging features to support those constructs.

Today, Visual Studio provides the following multithreaded debugging features for the modern .NET developer:

- **Threads**: This window shows a list of the threads that are used by your application while debugging. It also indicates which thread is active when it stopped at a breakpoint in your code.

- **Parallel Stacks**: This window allows developers to visualize the call stacks for each thread in their application in a single view. Selecting a thread in the window will display call stack information for the selected thread in the **Call Stack** window.

- **Parallel Watch**: This window works like the **Watch** window, except that you can see the value of a watch expression on each active thread in the application.

- **Debug Location**: This toolbar allows you to narrow your focus while debugging multithreaded applications. It has fields to select a **Process**, **Thread**, and **Stack Frame**. There are also buttons on the toolbar so that you can **Flag** and **Unflag** threads to be monitored.

- **Tasks**: This window displays each running task in the application and provides information about the thread that is running the task, the state of the task, and its call stack. You can also see the starting point for each task (the method or delegate that was passed to the task to be run).

- **Attach to Process**: This window allows you to attach the Visual Studio debugger to a process on the local machine or a remote machine. **Remote debugging** can be useful when working with multithreaded UI applications. It allows developers to debug their applications on systems with different numbers of processor cores than what's on their machines. They can also attach to a remote process running on a system running other processes that will be present in a production environment.

- **GPU Threads**: This window displays information about threads running on the GPU. This is used for C++ applications and is beyond the scope of this book. To learn more, you can read the documentation from Microsoft: `https://docs.microsoft.com/visualstudio/debugger/how-to-use-the-gpu-threads-window`.

In the sections ahead, we will use these debugging tools to step through multithreaded code in projects from some of the previous chapters of this book. Let's start by learning about the **Threads** and **Attach to Process** windows and the **Debug Location** toolbar.

Debugging threads and processes

In this section, we are going to debug **BackgroundPingConsoleApp** from *Chapter 1*. You can use your completed project from *Chapter 1* or get the project from this chapter's GitHub repository: `https://github.com/PacktPublishing/Parallel-Programming-and-Concurrency-with-C-sharp-10-and-.NET-6/tree/main/chapter10`. We will debug the application and discover some of the features of the **Debug Location** toolbar and the **Threads** window as we go.

Debugging a project with multiple threads

The project we'll be working this is a simple one that creates one background thread to check whether the network is available.

> **Note**
>
> The examples in this chapter will be run in the **Debug** configuration in Visual Studio. When you compile and run a .NET project, you can choose to run a **Debug** or **Release** build. While debugging, you will want to select **Debug** mode so that the project compiles w the symbolic debug information. This is not included in a **Release** build. For more information about build configurations, check out Microsoft Docs: `https://docs.microsoft.com/visualstudio/ide/understanding-build-configurations`.

Let's get started with our debugging example:

1. Start by opening **BackgroundPingConsoleApp** in Visual Studio and open `Program.cs` in the C# editor.

2. Set a breakpoint on the `Thread.Sleep(100)` statement inside the `while` loop.

3. Select **View** | **Toolbars** | **Debug Location** to display the **Debug Location** toolbar:

Figure 10.1 – The Debug Location toolbar in Visual Studio

We will be using this toolbar when we start debugging. All the fields are disabled when there is no active debugging session in Visual Studio.

4. Start debugging the project. When Visual Studio breaks at your breakpoint, notice the state of the **Debug Location** toolbar:

```
Process: [2040] BackgroundPingConsoleAp ▾    Lifecycle Events ▾  Thread: [9260] <No Name>    ▾        Stack Frame: Program.<Main>$.AnonymousMethod_( ▾

Program.cs ⊟ ×
C# BackgroundPingConsoleApp                                          ▾                                                              ▾
      1         Console.WriteLine("Hello, World!");
      2
      3    ⊟  var bgThread = new Thread(() =>
      4      {
      5    ⊟      while (true)
      6          {
      7              bool isNetworkUp = System.Net.NetworkInformation.NetworkInterface.GetIsNetworkAvailable();
      8              Console.WriteLine($"Is network available? Answer: {isNetworkUp}");
   ⟳  9              Thread.Sleep(100);
     10          }
     11      });
     12
     13      bgThread.IsBackground = true;
     14      bgThread.Start();
     15
     16  ⊟  for (int i = 0; i < 10; i++)
     17      {
     18          Console.WriteLine("Main thread working ... ");
     19          Task.Delay(500);
     20      }
     21
     22      Console.WriteLine("Done");
     23      Console.ReadKey();
     24
```

Figure 10.2 – Debugging with the Debug Location toolbar

The toolbar provides several dropdown controls to select the **Process**, **Thread**, and **Stack Frame** properties in scope. The **Process** dropdown will only contain a single process unless you explicitly debug multiple processes with the **Attach to Process** window. You can also set up multiple startup projects in Visual Studio to achieve this.

The **Threads** dropdown contains all the threads that belong to the selected process. The selected thread in this control is the background thread we created because the breakpoint was added within the code executed by that background thread.

The **Stack Frame** dropdown contains the list of frames in the current thread's call stack.

There is a **Toggle Current Thread Flagged State** button to the right of the **Threads** dropdown. We will learn about flagging threads later in the *Switching and flagging threads* section.

5. Next, select **Debug | Windows | Threads** to open the **Threads** window:

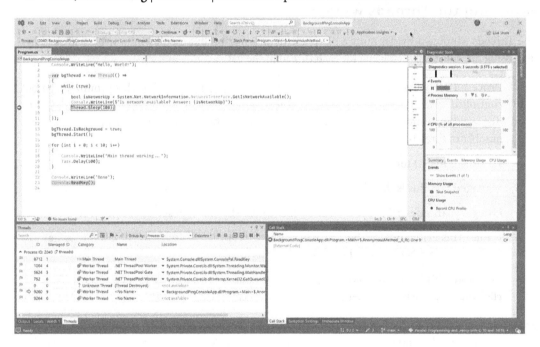

Figure 10.3 – Debugging with the Threads window active

By default, the **Threads** window will open in the lower-left panel with the **Output**, **Locals**, and **Watch** debugging windows.

6. Finally, expand the **Threads** window so that we can explore and discuss its features:

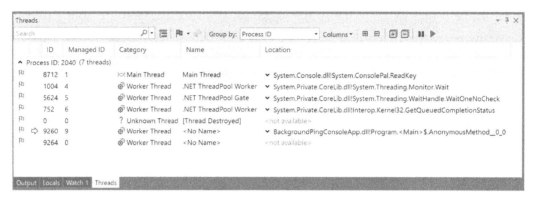

Figure 10.4 – Taking a closer look at the Threads window

Exploring the Threads window

The **Threads** window provides quite a bit of useful information in a small window. We will start by discussing the data that's displayed for each thread in the list:

- **Process ID**: By default, the list of threads is grouped by **Process ID**. This grouping can be controlled by the **Group by** dropdown in the window's toolbar. The **Process ID** grouping also displays the number of threads in its group. This can be useful when working with a large number of threads.

- **ID**: This is the ID for each thread in the list

- **Managed ID**: This is the `Thread.ManagedThreadId` property of each thread

- **Category**: This describes the type of thread (**Main Thread**, **Worker Thread**, and so on)

- **Name**: This field contains the `Thread.Name` property of each thread. If a thread has no name, then **<No Name>** will be displayed in this field.

- **Location**: This field contains the current stack frame of each thread in its call stack. You can click the dropdown in this field to display the full call stack for the thread.

Some additional fields are hidden by default. You can hide or show columns by selecting the **Columns** button in the **Threads** window toolbar. Select or unselect the columns you would like to show or hide. These are the columns that are hidden initially:

- **Priority**: This displays the priority assigned to the thread by the system

- **Affinity Mask**: The affinity mask determines which processors a thread can run on. This is determined by the system

- **Suspended Count**: This value is used by the system to decide whether the thread can be run

- **Process Name**: This is the name of the process that the thread belongs to

- **Process ID**: This is the ID of the process that the thread belongs to

- **Transport Qualifier**: This identifies the machine that is connected to the debugger. This is useful for remote debugging

Now, let's review the toolbar items available in the **Threads** window:

- **Search**: This allows you to search for threads. You can toggle the **Include call stacks in search** button on if you want the search results to encompass all call stack information

- **Flag**: With this dropdown button, you can select either **Flag Just My Code** or **Flag Custom Module Selection**

- **Group by**: This dropdown allows you to group threads by different fields. By default, they are grouped by **Process ID**

- **Columns**: This opens the **Columns** selection window so that you can customize the columns displayed in the **Threads** window

- **Expand/Collapse callstacks**: These two buttons expand or collapse the call stack in the **Location** column

- **Expand/Collapse groups**: These two buttons expand or collapse the thread groupings

- **Freeze Threads**: This freezes all selected threads in the window

- **Thaw Threads**: This unfreezes all selected threads in the window

Let's try the **Search** functionality. Start debugging the **BackgroundPingConsoleApp** project. When it hits the breakpoint, search for Anon in the **Search** field to find the thread whose call stack contains our anonymous method:

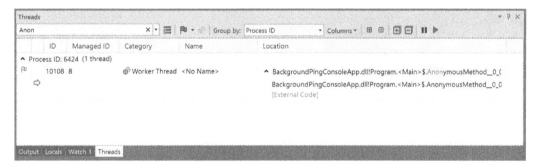

Figure 10.5 – Searching in the Threads window

The **Threads** window should now only contain the row for our **Worker Thread** with the **Anon** part of **AnonymousMethod** highlighted in *yellow*.

Now that you have some familiarity with the **Threads** window, let's learn how to use it to switch and flag threads.

Switching and flagging threads

The **Threads** window provides so much power when debugging a multithreaded application. We touched on some of these features in the previous section. In this section, we will learn how to switch threads, flag threads, and freeze or thaw a thread. Let's start by switching between threads in our **BackgroundPingConsoleApp** project.

Switching threads

You can switch context to a different thread by using the context menu in the **Threads** window. Run the project and wait for the debugger to pause at the breakpoint in our anonymous method. While the execution is paused in the debugger, right-click on the **Main Thread** row and select **Switch to Thread**. The cursor in the debugger should switch positions to the `Console.ReadLine()` statement. This is where the main thread is waiting for the user to press any key in the console:

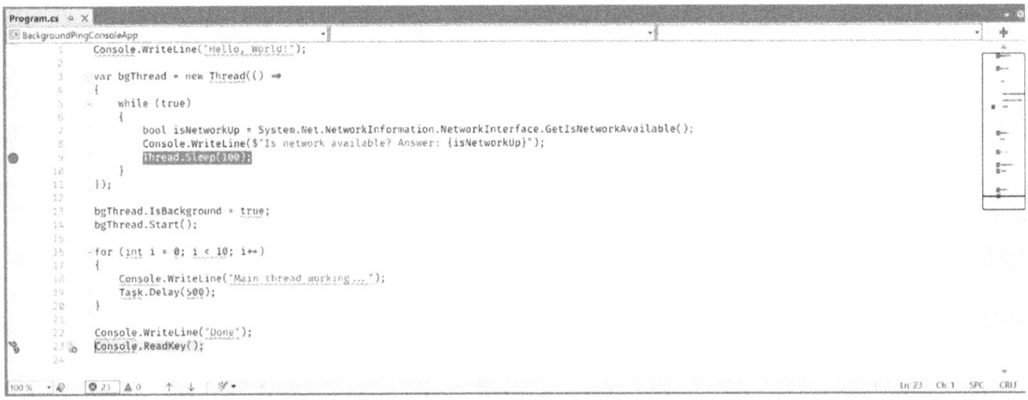

Figure 10.6 – Switching threads in the Visual Studio debugger

You can see how this function could be extremely useful when debugging a parallel operation with half a dozen active threads or more. Next, we will learn how to keep an eye on a specific thread with the **Flag Thread** feature.

Flagging threads

In this section, you will learn how to narrow your field of view while debugging in the **Threads** window. By only flagging the threads that we care about, we can reduce the clutter in the window. Here's how to flag threads:

1. If you aren't still debugging the **BackgroundPingConsoleApp** project, start debugging it now and wait for it to stop at the breakpoint.

2. While the debugger is paused in the application, right-click the **Main Thread** row and select **Flag**. The flag icon should now be colored *orange* in that row.

3. Do the same for the row containing **Worker Thread** with **AnonymousMethod** in the call stack

4. Next, click the **Show Flagged Threads Only** button in the window's toolbar:

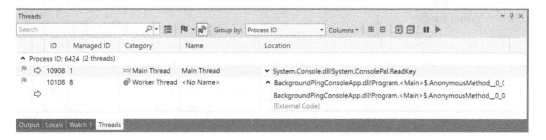

Figure 10.7 – Showing flagged threads only in the Threads window

This makes it simpler to track only the threads that are important to our current debugging session. You can click the button again to toggle the button off and view all threads. It is also possible to flag threads in the **Parallel Watch** and **Parallel Stacks** windows. Their flagged state will persist across all of these windows and the **Debug Location** toolbar.

There's an even easier way to flag these two threads in our application. These are the only two threads that are part of our application's code. So, we can use the **Flag Just My Code** button in the toolbar to flag them.

1. Unselect the **Show Flagged Threads Only** toolbar button

2. Right-click one of the flagged rows in the window and select **Unflag All**

3. Now, click **Flag Just My Code** in the toolbar. The same two threads will be flagged again:

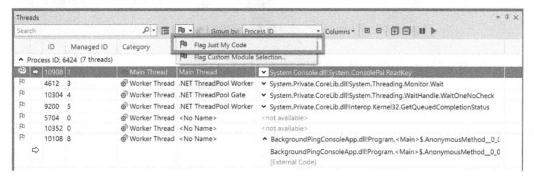

Figure 10.8 – Flagging only the threads that belong to our code

This is much easier than selecting threads one by one in the list. It may not always be as obvious which threads are part of our code. In the next section, we will learn how to freeze a thread.

Freezing threads

Freezing or thawing a thread in the **Threads** window is the equivalent of calling the `SuspendThread` or `ResumeThread` Windows functions. If a frozen thread is not executing any code yet, it will never start unless it is thawed. If a thread is currently executing, it will pause when the **Freeze** thread is called in Visual Studio.

Let's try freezing and thawing the worker thread in our **BackgroundPingConsoleApp** project to see what happens in the debugger:

1. Before running the application, add new breakpoints at the `while (true)` and `Console. ReadKey()` statements. Keep the existing breakpoint at `Thread.Sleep(100)`

2. Start debugging the application

3. When the debugger breaks on the `while (true)` line, right-click the worker thread that contains **AnonymousMethod** and select **Freeze**

4. Continue debugging; it should break on the `Console.ReadKey()` line instead of `Thread. Sleep(100)`. This is because the worker thread is not currently running:

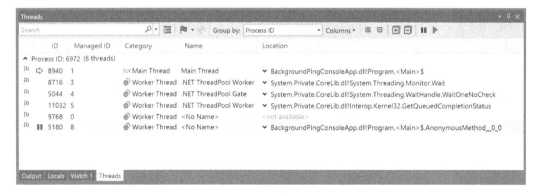

Figure 10.9 – Freezing a worker thread in the Threads window

5. Right-click the worker thread again and select **Thaw**

6. Now, continue debugging again. Visual Studio breaks on the `Thread.Sleep(100)` line inside the anonymous method.

This shows how the functions of the **Threads** window could be extremely useful while debugging a multithreaded application.

Now that we have learned how to debug our multithreaded application by switching, freezing, and flagging threads with the **Threads** window, let's learn how we can leverage additional features such as the **Parallel Stacks** and **Parallel Watch** windows while debugging.

Debugging a parallel application

Visual Studio provides several windows for parallel debugging. While the **Threads** window excels for any type of multithreaded application, other windows provide additional features and views when working with Task objects in our applications.

We will start our tour of these features with the **Parallel Stacks** window.

Using the Parallel Stacks window

The **Parallel Stacks** window provides a visual representation of the threads or tasks in the application. These are two distinct views in the window. You can switch between them by selecting **Threads** or **Tasks** in the **View** dropdown box. The following screenshot shows an example of the **Threads** view while debugging the **BackgroundPingConsoleApp** project:

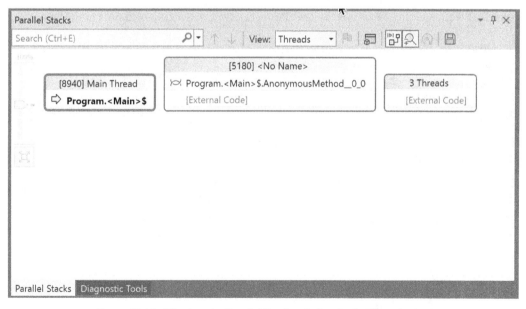

Figure 10.10 – Viewing the Parallel Stacks window in the Threads view

The **Parallel Stacks** window contains a toolbar with the following items from left to right. You can follow along by examining the tooltips for the toolbar items in the window in Visual Studio:

- **Search**: This allows the same type of search functionality that is available in the **Threads** window. It has the **Find Previous** and **Find Next** buttons to the right of the **Search** field.

- **View**: This dropdown switches between the **Threads** and **Tasks** views

- **Show Only Flagged**: This toggle will hide any threads that are not flagged

- **Toggle Method View**: This will switch to a view of the currently selected method and its call stack

- **Auto Scroll to Current Stack Frame**: This will scroll the current stack frame into view in the diagram while stepping through the debugger. This option is toggled on by default.

- **Toggle Zoom Control**: This hides or shows the zoom control on the diagram's surface. This option is turned on by default.

- **Reverse Layout**: This option mirrors the layout of the current view

- **Save Diagram**: This option saves the current diagram to a `.png` file

To examine the **Tasks** view of the window, we will need to open a different project that has some `Task` objects. Let's work with the **Tasks** view by opening a project from a previous chapter in the book:

1. Open your **TaskSamples** project from *Chapter 5*, or get a copy of this project from this chapter's source code on GitHub: `https://github.com/PacktPublishing/Parallel-Programming-and-Concurrency-with-C-sharp-10-and-.NET-6/tree/main/chapter10`.

2. Open `Examples.cs` and set a breakpoint on the first line of the `ProcessOrders` method.

3. Start debugging. When the debugger stops on the breakpoint, select **Debug | Windows | Parallel Stacks**.

4. Switch to the **Tasks** view in the **Parallel Stacks** window:

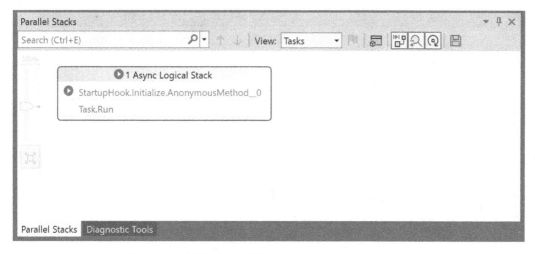

Figure 10.11 – The Parallel Stacks window in the Tasks view

No tasks have been started yet, so there isn't much to see here. There is a single **Async Logical Stack** block that looks like it is ready to start analyzing some async work.

5. Add a breakpoint on the `Tasks.WaitAll` statement and click **Continue**

Note

It is possible to configure breakpoints in Visual Studio by right-clicking on the breakpoint you want to modify and clicking **Settings**. If you select **Filter** under **Conditions** in the breakpoint settings, you can add a filter based on one or more `ThreadId` or `ThreadName` values. This will ensure that the debugger will only stop on the current breakpoint when the desired thread(s) are executing that line of code. To read more about breakpoint conditions and filters, check out this article on Microsoft Docs: `https://docs.microsoft.com/visualstudio/debugger/using-breakpoints#set-a-filter-condition`.

---

6. Now, examine the **Parallel Stacks** window again:

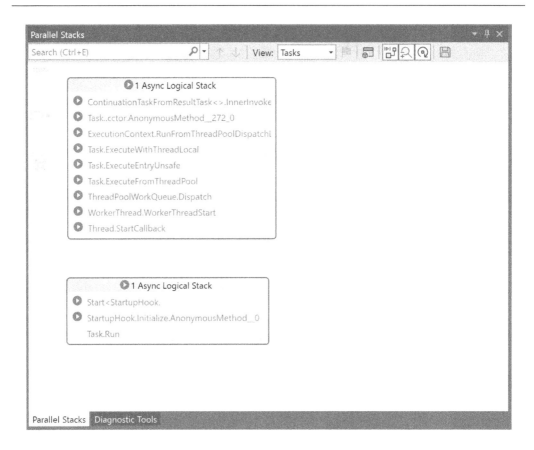

Figure 10.12 – The Parallel Stacks window while tasks are active

> **Note**
> It can be challenging to catch the tasks in this window while they are still executing if they are fast-running methods. You may have to run the application several times to hit this breakpoint if one or more of the Task objects has not been completed yet.

In this case, the **Parallel Stacks** window has captured the execution of one running task and another preparing to run. There are some differences between this **Tasks** view and some of the thread analysis we have done in this chapter:

- Only actively running tasks are shown in the **Tasks** view
- The **Tasks** view's stack attempts to display only the relevant call stack information. Stack frames may be trimmed from the top and bottom if they are not relevant. If you need to see the entire call stack, switch back to the **Threads** view.

- A separate block is displayed for each active task in the **Tasks** view, even if they are assigned to the same thread.

You can hover over a line in a task's call stack to view more information about its thread and stack frame:

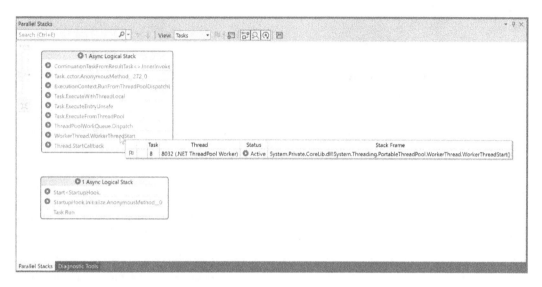

Figure 12.13 – Viewing more information about a call stack frame

If you want to pivot the **Tasks** view to a particular method, you can use the **Toggle Method View** button:

1. Start a new debugging session in the **TaskSamples** project
2. Set a new breakpoint on the `return orders` statement in the `PrepareOrders` method
3. Click **Continue**. The **Parallel Watch** window will display the active tasks when the debugger breaks inside the `PrepareOrders` method.
4. Click the **Toggle Method View** button. You now have a method-focused view of the **Tasks** view and can hover over the `PrepareOrders` method to get more call stack and thread information:

Figure 10.14 – Leveraging the Method View area of the Parallel Stacks window

Next, we will learn how to view the state of variables across different threads by using the **Parallel Watch** window.

Using the Parallel Watch window

The **Parallel Watch** window is similar to the **Watch** window in Visual Studio, but it displays additional information about the value of the watched expression across the threads with access to the data in the expression.

In this example, we will modify the Examples class in the **TaskSamples** project to add a state that will be available to multiple threads:

1. Start by adding a private variable to the Examples class:

    ```
    private List<Order> _sharedOrders;
    ```

2. Add a line to ProcessOrders to assign orders to _sharedOrders:

    ```
    private List<Order> PrepareOrders(List<Order> orders)
    {
        // TODO: Prepare orders here
        _sharedOrders = orders;
        return orders;
    }
    ```

3. Keep the breakpoints from the previous example and start debugging. Continue until the debugger breaks on the return orders statement inside ProcessOrders.

4. Select **Debug** | **Windows** | **Parallel Watch 1** to open the **Parallel Watch 1** window. You can open up to four **Parallel Watch** windows to separate your watched expressions.

5. In the **Parallel Watch 1** window, you will see a line for the current thread in context. Add a watch to the _sharedOrders private variable:

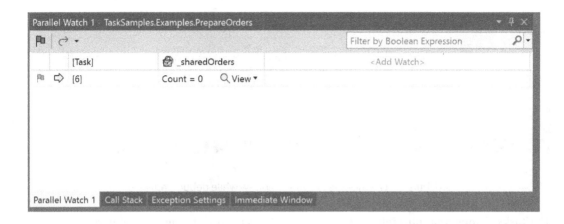

Figure 10.15 – Adding a watched expression in Parallel Watch 1 window

The window indicates that **Task 6** has _sharedOrders in scope and that the count of orders in the variable is 0.

6. Right-click on **Main Thread** in the **Threads** window and select **Switch to Thread**. In the **Parallel Watch 1** window, a task is no longer in scope, so the header label has changed from **Task** to **Thread**, and the **ID** property of **Main Thread** will be displayed:

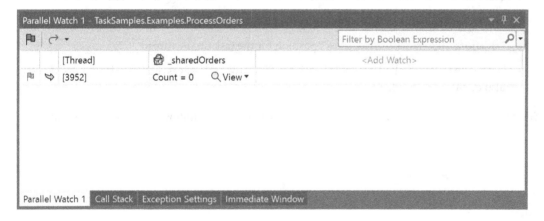

Figure 10.16 – Viewing the watched variable on Main Thread

7. Finally, select **Debug | Windows | Tasks** to open the **Tasks** window:

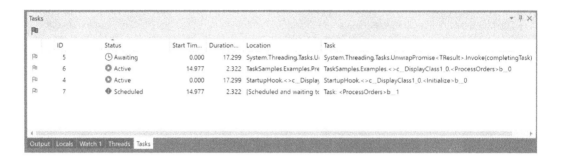

Figure 10.17 – Viewing the Tasks window while debugging

The **Tasks** window will show information about the tasks in scope in the debugging session. The following columns are displayed in the window:

- **Flag**: An icon indicating whether the current task has been flagged. You can click this field to flag or unflag a task.

- **ID**: The ID of the task

- **Status**: The `Task.Status` properties of the task

- **Start Time (sec)**: This indicates how many seconds into the debugging session the task started

- **Duration (sec)**: This indicates how long the task has been running

- **Location**: This shows the call stack's position for the task on the thread

- **Task**: The initial method where the task started. Any parameters that have been passed will also be shown in this field.

Several other hidden fields can be shown by right-clicking in the window and selecting **Columns**:

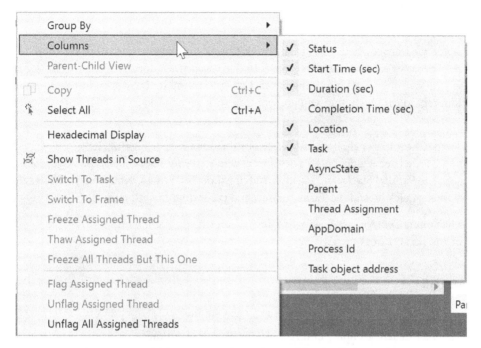

Figure 10.18 – Adding or removing columns from the Tasks window

You can sort and group the tasks in the **Tasks** window similar to how the **Threads** window works. The difference is that the **Tasks** window does not have a toolbar. All operations are performed with the right-click context menu.

The other tool you can use while debugging parallel .NET code is the **Debug Location** toolbar. If it is not already displayed in Visual Studio, you can open it by going to **View | Toolbars | Debug Location**. While you're debugging, the toolbar functionality lights up:

Figure 10.19 – Viewing the Debug Location toolbar while debugging

From the toolbar, you can select the active **Process**, **Thread**, and **Stack Frame**. It's also easy to toggle the flagged state of the currently selected thread.

That completes our tour of the debug windows available to .NET parallel programmers. Let's wrap up by reviewing everything we learned in this chapter.

Summary

In this chapter, we learned about the Visual Studio features available to multithreaded application developers. We started by working with threads in the **Threads** window. This is the most universal debugging window when working doing parallel programming in .NET. It can provide essential information, regardless of whether you are working with async tasks, parallel loops, or standard `Thread` objects.

Next, we learned how to switch, flag, and freeze our threads while debugging. Finally, we looked at some of the advanced debugging tools for developers who are using `Task` objects or `async/await` in their code. The **Parallel Stacks** and **Parallel Watch** windows take task debugging to the next level. Finally, we took a quick look at the **Tasks** window and the **Debug Location** toolbar.

In the next chapter, *Chapter 11*, we will dive deeper into the different methods available to cancel concurrent and parallel work with .NET.

Questions

1. How can you debug multiple processes in Visual Studio?
2. What is the default grouping of threads in the **Threads** window?
3. How can you add more columns to the **Tasks** or **Threads** window?
4. Which debug window displays a visual representation of the current threads or tasks?
5. What file format can you export from the **Parallel Stacks** window?
6. How many **Parallel Watch** windows can you open?
7. Which Visual Studio toolbar provides information about the processes and threads you are currently debugging?
8. How can you filter the **Threads** window to only show the threads that have been created for your code?

Canceling Asynchronous Work

In the previous chapters, we've looked at a few examples of how to cancel threads and tasks. This chapter will explore more of the methods available to cancel concurrent and parallel work with C# and .NET. The methods in this chapter will provide alternative ways to cancel background operations using callbacks, polling, and wait handles. You will gain a deeper understanding of how to safely cancel asynchronous work with a variety of methods using some practical scenarios.

In this chapter, you will learn about the following topics:

- Canceling managed threads

- Canceling parallel work

- Discovering patterns for thread cancellation

- Handling multiple cancelation sources

By the end of this chapter, you will understand how to cancel different types of asynchronous and parallel tasks.

Technical requirements

To follow along with the examples in this chapter, the following software is recommended for Windows developers:

- Visual Studio 2022 version 17.2 or later.

- .NET 6.

- To complete any WinForms or WPF samples, you will need to install the .NET desktop development workload for Visual Studio. These projects will only run on Windows.

All the code examples for this chapter can be found on GitHub at `https://github.com/PacktPublishing/Parallel-Programming-and-Concurrency-with-C-sharp-10-and-.NET-6/tree/main/chapter11`.

Canceling managed threads

Canceling asynchronous work in .NET is based on the use of a **cancellation token**. A token is a simple object that is used to signal that a cancellation request has been made to another thread. The `CancellationTokenSource` object manages these requests and contains a token. If you want to cancel several operations with the same trigger, the same token should be provided to all of the threads to be canceled.

A `CancellationTokenSource` instance has a `Token` property to access the `CancellationToken` property and pass it to one or more asynchronous operations. The request to cancel can only be made from the `CancellationTokenSource` object. The `CancellationToken` property provided to the other operations receives the signal to cancel but cannot initiate a cancellation.

`CancellationTokenSource` implements the `IDisposable` interface, so be sure to call `Dispose` when you are freeing your managed resources. A `using` statement or block to automatically dispose of the token source would be preferred if it is practical for your workflow.

It is important to understand that cancellation is not forced on the listening code. The asynchronous code that receives a cancellation request must determine whether it can currently cancel its work. It might decide to immediately cancel, cancel after finishing some intermediate tasks, or finish its work and ignore the request. There can be valid reasons why a routine will ignore a request to cancel. It is possible that the work is almost complete or that canceling in the current state will cause some data corruption. The decision to cancel must be mutual between the requestor and the listener.

Let's look at an example of how to cooperatively cancel some work being processed on a background thread on the `ThreadPool` thread:

1. In Visual Studio, create a new .NET 6 console application named `CancelThreadsConsoleApp`.

2. Add a new class named `ManagedThreadsExample`.

3. Create a method named `ProcessText` in the `ManagedThreadsExample` class:

    ```
    public static void ProcessText(object? cancelToken)
    {
        var token = cancelToken as CancellationToken?;
        string text = "";

        for (int x = 0; x < 75000; x++)
    ```

```
        {
            if (token != null && token.Value
                .IsCancellationRequested)
            {
                Console.WriteLine($"Cancellation request
                    received. String value: {text}");
                break;
            }
            text += x + " ";
            Thread.Sleep(500);
        }
    }
```

This method appends the value of the iterator variable, x, to the `string` variable of `text` until a cancellation request is received. There is a `Thread.Sleep(500)` statement to allow the calling method some time to cancel the operation.

4. Next, create a method named `CancelThread`, in `Program.cs`:

```
private static void CancelThread()
{
    using CancellationTokenSource tokenSource = new();
    Console.WriteLine("Starting operation.");
    ThreadPool.QueueUserWorkItem(new
        WaitCallback(ManagedThreadsExample
            .ProcessText), tokenSource.Token);
    Thread.Sleep(5000);
    Console.WriteLine("Requesting cancellation.");
    tokenSource.Cancel();
    Console.WriteLine("Cancellation requested.");
}
```

This method calls `ThreadPool.QueueUserWorkItem` to queue the `ProcessText` method in the `ThreadPool` thread. The method also receives a cancellation token from `tokenSource.Token`. After waiting for five seconds, `tokenSource.Cancel` is called, and `ProcessText` will receive the cancellation request.

Notice that `tokenSource` is created in a `using` statement. This ensures that it will be properly disposed of when it goes out of scope.

5. Add a call to `CancelThread` to the `Main` method in `Program.cs`:

    ```
    static void Main(string[] args)
    {
        CancelThread();
        Console.ReadKey();
    }
    ```

6. Finally, run the application and observe the console output:

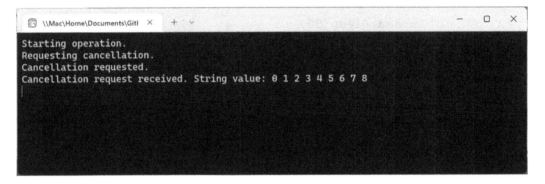

Figure 11.1 – Running the CancelThreadsConsoleApp project

The `for` loop should have enough time to execute 9 or 10 times before receiving the request to cancel. How does your output match up?

Now that we have covered some basics of cancellation and worked with a common method of using a cancellation token, let's create some examples of how to cancel parallel loops and PLINQ queries.

Canceling parallel work

In this section, we will work with some examples of canceling parallel operations. There are a few operations that fall into this realm. There are static parallel operations that are part of the `System.Threading.Tasks.Parallel` class and there are PLINQ operations. Both of these types use a `CancellationToken` property, as we used in our managed threading example in the previous section. However, handling the cancellation request is slightly different. Let's look at an example to understand the differences.

Canceling a parallel loop

In this section, we will create a sample that illustrates how to cancel a `Parallel.For` loop. The same method of cancellation is leveraged for the `Parallel.ForEach` method. Perform the following steps:

1. Open the `CancelThreadsConsoleApp` project from the previous section.

2. In the `ManagedThreadsExample` class, create a new `ProcessTextParallel` method with the following implementation:

```
public static void ProcessTextParallel(object?
    cancelToken)
{
    var token = cancelToken as CancellationToken?;
    if (token == null) return;
    string text = "";
    ParallelOptions options = new()
    {
        CancellationToken = token.Value,
        MaxDegreeOfParallelism =
            Environment.ProcessorCount
    };
    try
    {
        Parallel.For(0, 75000, options, (x) =>
        {
            text += x + " ";
            Thread.Sleep(500);
        });
    }
    catch (OperationCanceledException e)
    {
        Console.WriteLine($"Text value: {text}.
            {Environment.NewLine} Exception
                encountered: {e.Message}");
    }
}
```

Essentially, the preceding code does the same thing as the `ProcessText` method in our last example. It appends a numeric value to the `text` variable until a cancellation is requested. Let's examine the differences:

- First, we are setting `token.Value` to the `CancellationToken` property of a `ParallelOptions` object. These options are passed as the third parameter to the `Parallel.For` method.

- The second major difference is that we handle the cancellation request by catching an `OperationCanceledException` type. This exception type will be thrown when our other code in `Program.cs` requests a cancellation.

3. Next, add a method named `CancelParallelFor` to `Program.cs`:

```
private static void CancelParallelFor()
{
    using CancellationTokenSource tokenSource = new();
    Console.WriteLine("Press a key to start, then
        press 'x' to send cancellation.");
    Console.ReadKey();
    Task.Run(() =>
    {
        if (Console.ReadKey().KeyChar == 'x')
            tokenSource.Cancel();
        Console.WriteLine();
        Console.WriteLine("press a key");
    });
    ManagedThreadsExample.ProcessTextParallel
        (tokenSource.Token);
}
```

In this method, the user is instructed to press a key to start the operation and to press the *X* key when they are ready to cancel the operation. The code to handle receiving x `KeyChar` from the console and sending a `Cancel` request is performed on another thread in order to keep the current thread free to call `ProcessTextParallel`.

4. Finally, update the `Main` method to call `CancelParallelFor` and comment out the call to `CancelThread`:

```
static void Main(string[] args)
{
    //CancelThread();
```

```
            CancelParallelFor();
            Console.ReadKey();
        }
```

5. Now run the project. Follow the prompts to cancel the `Parallel.For` loop, and examine the output:

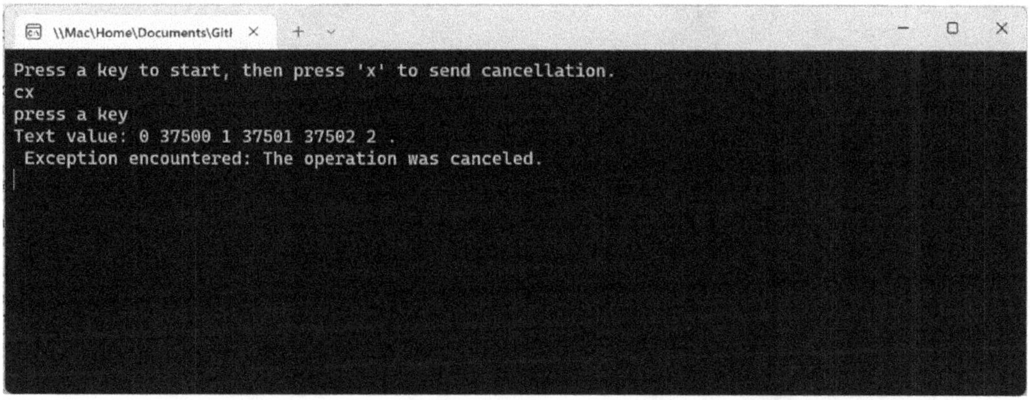

Figure 11.2 – Canceling a Parallel.For loop from the console

> Notice how the numbers are not in sequence at all. In this case, it appears that the `Parallel.For` operation used two different threads. The first thread started at 0, while the second thread was operating on integers starting with 37500. This is the midway point of the maximum value of 75000 provided to the method parameter.

In the next section, we will briefly examine how to cancel a PLINQ query.

Canceling a PLINQ query

Canceling a PLINQ query is also achieved by catching the `OperationCanceledException` type. However, instead of using the `ParallelOptions` object that is used with parallel loops, you can call `WithCancellation` as part of the query.

To learn how to cancel a PLINQ query, let's walk through an example:

1. Start this example by adding a method named `ProcessNumsPlinq`, to the `ManagedThreadsExample` class:

```
    public static void ProcessNumsPlinq(object?
        cancelToken)
    {
        int[] input = Enumerable.Range(1,
```

```
            25000000).ToArray();
        var token = cancelToken as CancellationToken?;
        if (token == null) return;
        int[]? result = null;
        try
        {
            result =
                (from value in input.AsParallel()
                    .WithCancellation(token.Value)
                    where value % 7 == 0
                    orderby value
                    select value).ToArray();
        }
        catch (OperationCanceledException e)
        {
            Console.WriteLine($"Exception encountered:
                {e.Message}");
        }
    }
```

This method creates an array of 25 million integers and uses the PLINQ query to determine which of them are divisible by seven. The token.Value is passed to the WithCancellation operation in the query. When an exception is thrown by a cancellation request, the exception details are written to the console.

2. Next, add a method named CancelPlinq to Program.cs:

```
private static void CancelPlinq()
{
    using CancellationTokenSource tokenSource = new();
    Console.WriteLine("Press a key to start.");
    Console.ReadKey();
    Task.Run(() =>
    {
        Thread.Sleep(100);
        Console.WriteLine("Requesting cancel.");
        tokenSource.Cancel();
        Console.WriteLine("Cancel requested.");
    });
```

```
ManagedThreadsExample.ProcessNumsPlinq
    (tokenSource.Token);
}
```

This time, the cancellation will be triggered automatically after 100 milliseconds.

3.   Update the `Main` method to call `CancelPlinq`, and run the application:

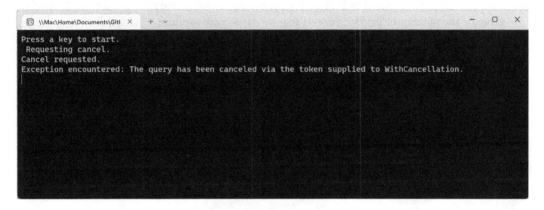

```
Press a key to start.
 Requesting cancel.
Cancel requested.
Exception encountered: The query has been canceled via the token supplied to WithCancellation.
```

Figure 11.3 – Canceling a PLINQ operation in the console application

Unlike the previous examples, there is no query output to examine. You cannot get partial output from a PLINQ query. The `result` variable will be `null`.

In the next section, we will work with some different methods of cancellation.

Discovering patterns for thread cancellation

There are different methods of listening for cancellation requests from a thread or task. So far, we have seen examples of managing these requests by either handling the `OperationCanceledException` type or checking the value of `IsCancellationRequested`. The pattern of checking `IsCancellationRequested`, usually inside a loop, is called **polling**. First, we will see another example of this pattern. The second pattern we will examine is receiving the notification by **registering a callback method**. The final pattern that we will cover in this section is **listening to cancellation requests with wait handles** using `ManualResetEvent` or `ManualResetEventSlim`.

Let's start by trying another example of handling a cancellation request by polling.

Canceling with polling

In this section, we will create another example that uses polling to cancel a background task. The previous example of polling was running in a background thread on the `ThreadPool` thread. This example will also start a `ThreadPool` thread, but it will leverage `Task.Run` to start the background thread. We will create and process a million System.Drawing.Point objects, finding those with a Point.X value of less than 50. Users will have the option to cancel processing by pressing the *X* key:

1. Start by creating a new .NET console application project named CancellationPatterns

2. Add a new class to the project named PollingExample

3. Add a private static method to PollingExample named GeneratePoints. This will generate the number of Point objects that we desire with random X values:

```
private static List<Point> GeneratePoints(int count)
{
    var rand = new Random();
    var points = new List<Point>();
    for (int i = 0; i <= count; i++)
    {
        points.Add(new Point(rand.Next(1, count * 2),
            100));
    }
    return points;
}
```

4. Don't forget to add a using statement to use the Point type:

```
using System.Drawing;
```

5. Next, add a private static method named FindSmallXValues to PollingExample. This method loops through the list of points and outputs those with an X value of less than 50. Each time through the loop, it checks the token for cancellation and breaks out of the loop when it occurs:

```
private static void FindSmallXValues(List<Point>
    points, CancellationToken token)
{
    foreach (Point point in points)
    {
        if (point.X < 50)
        {
            Console.WriteLine($"Point with small X
                coordinate found. Value: {point.X}");
        }
    }
```

```
        if (token.IsCancellationRequested)
        {
            break;
        }
        Thread.SpinWait(5000);
    }
}
```

A `Thread.SpinWait` statement is added at the end of the loop to give users some time to cancel the operation.

6. Add a public static method to `PollingExample` named `CancelWithPolling`:

```
public static void CancelWithPolling()
{
    using CancellationTokenSource tokenSource = new();
    Task.Run(() => FindSmallXValues(GeneratePoints
        (1000000), tokenSource.Token), tokenSource
            .Token);
    if (Console.ReadKey(true).KeyChar == 'x')
    {
        tokenSource.Cancel();
        Console.WriteLine("Press a key to quit");
    }
}
```

The preceding method creates the `CancellationTokenSource` object and passes it to `FindSmallXValues` and also `Task.Run`. If you wanted to cancel the `Task`, instead of breaking out of the loop when `IsCancellationRequested` becomes `true`, you would call `token.ThrowIfCancellationRequested`. This would throw an exception in the `Task`. The `CancelWithPolling` method would then require a `try/catch` block around the `Task.Run` call. It's a best practice to use exception handling with all multithreaded code anyway. In this case, you would have two exception handlers: one to handle `OperationCanceledException` and a second to handle `AggregateException`.

Additionally, the `CancelWithPolling` method has code to determine when the user presses the *X* key to cancel the operation.

7. Finally, open `Program.cs` and add some code to execute the sample:

```
using CancellationPatterns;
Console.WriteLine("Hello, World! Press a key to start,
    then press 'x' to cancel.");
Console.ReadKey();
PollingExample.CancelWithPolling();
Console.ReadKey();
```

8. Now run the application, and examine the output:

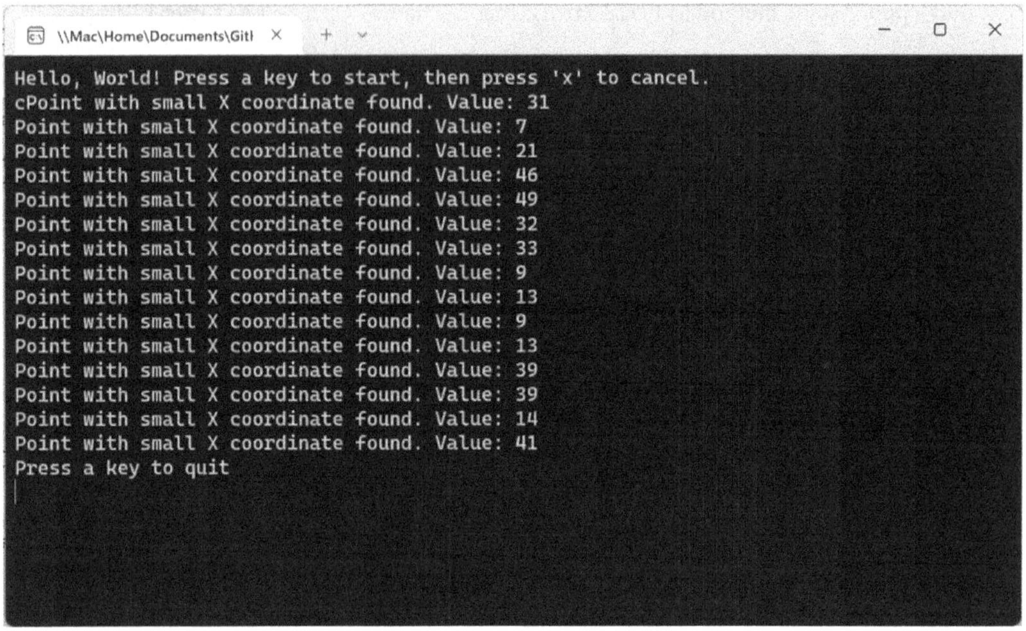

Figure 11.4 – Running the cancellation polling example

Depending on how long you wait before canceling, you might have a different number of points found by the process.

In the next section, we will learn how you can register a callback method to handle cancellation requests.

Canceling with callbacks

Some code in .NET supports the registration of a callback method to cancel processing. One class that supports cancellation with callbacks is `System.Net.WebClient`. In this example, we will use `WebClient` to start downloading a file. The download will be canceled after three seconds. To ensure the file download is large enough that it has not been completed after three seconds, we will download a large lossless audiobook file from **Internet Archive** (`https://archive.org/`). We will download the first part of the audiobook of *The Odyssey* by Homer. This file is 471.1 MB. You can view all of the free downloads for this book at `https://archive.org/details/lp_the-odyssey_homer-anthony-quayle`. Perform the following steps:

1. Open the **CancellationPatterns** project and add a new class named `CallbackExample`

2. Start by adding a method named `GetDownloadFileName` to build the path where the file will be downloaded. We will download it to the same folder where our assembly is executing:

```
private static string GetDownloadFileName()
{
    string path = System.Reflection.Assembly
        .GetAssembly(typeof(CallbackExample)).Location;
    string folder = Path.GetDirectoryName(path);
    return Path.Combine(folder, "audio.flac");
}
```

3. Next, add an `async` method named `DownloadAudioAsync`. This method will handle the file download and cancellation. There are several exception handlers to catch any type of exception that the `DownloadFileTaskAsync` method might throw. In turn, all of them throw an `OperationCanceledException` type to be handled by the parent method:

```
private static async Task DownloadAudioAsync
    (CancellationToken token)
{
    const string url = "https://archive.org/download/
        lp_the-odyssey_homer-anthony-quayle/disc1/
            lp_the-odyssey_homer-anthony-quayle
                _disc1side1.flac";
    using WebClient webClient = new();
    token.Register(webClient.CancelAsync);
    try
    {
        await webClient.DownloadFileTaskAsync(url,
            GetDownloadFileName());
    }
    catch (WebException we)
    {
```

```
            if (we.Status == WebExceptionStatus
                .RequestCanceled)
                throw new OperationCanceledException();
        }
        catch (AggregateException ae)
        {
            foreach (Exception ex in ae.InnerExceptions)
            {
                if (ex is WebException exWeb &&
                    exWeb.Status == WebExceptionStatus
                        .RequestCanceled)
                    throw new OperationCanceled
                        Exception();
            }
        }
        catch (TaskCanceledException)
        {
            throw new OperationCanceledException();
        }
    }
```

4. Add a using statement for the WebClient type:

    ```
    using System.Net;
    ```

5. Now add a public async method named CancelWithCallback. This method
    calls DownloadAudioAsync, waits for three seconds, and calls Cancel on the
    CancellationTokenSource object. Awaiting the task in a try block means we can
    handle the OperationCanceledException type directly. If you used task.Wait
    instead, you would have to catch AggregateException and check whether one of the
    InnerException objects is an OperationCanceledException type:

    ```
    public static async Task CancelWithCallback()
    {
        using CancellationTokenSource tokenSource = new();
        Console.WriteLine("Starting download");
        var task = DownloadAudioAsync(tokenSource.Token);
        tokenSource.Token.WaitHandle.WaitOne
            (TimeSpan.FromSeconds(3));
    ```

```
        tokenSource.Cancel();
        try
        {
            await task;
        }
        catch (OperationCanceledException ex)
        {
            Console.WriteLine($"Download canceled.
                Exception: {ex.Message}");
        }
    }
```

In this step, it might be necessary to adjust the number of seconds in the `tokenSource.Token.WaitHandle.WaitOne` call. The timing can vary based on your computer's download speed and processing speed. Try adjusting the value if you do not see a `Download canceled` message in the console output.

6. Finally, comment out the existing code in `Program.cs`, and add the following code to call the `CallbackExample` class:

```
using CancellationPatterns;
await CallbackExample.CancelWithCallback();
Console.ReadKey();
```

7. Now run the application, and examine the output:

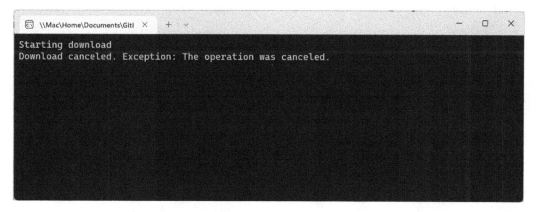

Figure 11.5 – Canceling a download with CancellationToken and a callback

You can verify that the download started and did not complete by looking in the folder where your assembly was running. You should see a file named **audio.flac** with a file size of **0 KB**. You can safely delete this file as it could cause exceptions if you try to download it again.

Now that we have seen how to cancel a background task with a callback method, let's wrap up this section by working through an example with wait handles.

Canceling with wait handles

In this section, we will use `ManualResetEventSlim` to cancel a background task that would not otherwise be responsive to user input. This object has `Set` and `Reset` events to start/ resume or pause an operation. When operations have not yet started or have been paused, calling `ManualResetEventSlim.Wait` will cause the operation to pause on that statement until another thread calls `Set` to start or resume processing.

This example will iterate over 100,000 integers and output to the console for each even number. This process can be started, paused, resumed, or canceled thanks to the `ManualResetEventSlim` object and `CancellationToken`. Let's try this example in our project:

1. Start by adding a `WaitHandleExample` class to the **CancellationPatterns** project.

2. Add a private variable named `resetEvent` to the new class:

   ```
   private static ManualResetEventSlim resetEvent =
       new(false);
   ```

3. Add a private static method named `ProcessNumbers` to the class. This method iterates over the numbers and only continues processing when `resetEvent.Wait` allows it to proceed:

   ```
   private static void ProcessNumbers(IEnumerable<int>
       numbers, CancellationToken token)
   {
       foreach (var number in numbers)
       {
           if (token.IsCancellationRequested)
           {
               Console.WriteLine("Cancel requested");
               token.ThrowIfCancellationRequested();
           }
           try
           {
               resetEvent.Wait(token);
   ```

```
        }
        catch (OperationCanceledException)
        {
            Console.WriteLine("Operation canceled.");
            break;
        }
        if (number % 2 == 0)
            Console.WriteLine($"Found even number:
                {number}");
        Thread.Sleep(500);
    }
}
```

4. Next, add a public static async method named `CancelWithResetEvent` to the class. This method creates the list of numbers to process, calls `ProcessNumbers` within a `Task.Run` call, and uses a `while` loop to listen for user input:

```
public static async Task CancelWithResetEvent()
{
    using CancellationTokenSource tokenSource = new();
    var numbers = Enumerable.Range(0, 100000);
    _ = Task.Run(() => ProcessNumbers(numbers,
        tokenSource.Token), tokenSource.Token);
    Console.WriteLine("Use x to cancel, p to pause, or
        s to start or resume,");
    Console.WriteLine("Use any other key to quit the
        program.");
    bool running = true;
    while (running)
    {
        char key = Console.ReadKey(true).KeyChar;
        switch (key)
        {
            case 'x':
                tokenSource.Cancel();
                break;
            case 'p':
```

```
                    resetEvent.Reset();
                    break;
                case 's':
                    resetEvent.Set();
                    break;
                default:
                    running = false;
                    break;
            }
            await Task.Delay(100);
        }
    }
```

5. Finally, update `Program.cs` to contain the following code:

```
using CancellationPatterns;
await WaitHandleExample.CancelWithResetEvent();
Console.ReadKey();
```

6. Run the program to test it. Follow the console prompts to start, pause, resume, and cancel the process:

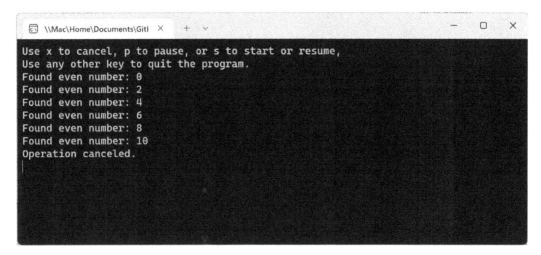

Figure 11.6 – Testing the CancelWithResetEvent method in the console

You should see in the console output that several event numbers have been found before the operation is canceled. The amount of processing completed could vary based on your computer's processors.

In the next section, we will wrap up cancellation by learning how to handle cancellation requests from multiple sources.

Handling multiple cancellation sources

Background tasks can leverage `CancellationTokenSource` to receive cancellation requests from as many sources as necessary. The static `CancellationTokenSource.CreateLinkedTokenSource` method accepts an array of `CancellationToken` objects to create a new `CancellationTokenSource` object that will notify us of cancellation if any of the source tokens receives a request to cancel.

Let's look at a quick example of how to implement this in our **CancellationPatterns** project:

1. First, open the `PollingExample` class. We are going to create an overload of the `CancelWithPolling` method that accepts a `CancellationTokenSource` parameter. The two overloads of `CancelWithPolling` will look like this:

```
public static void CancelWithPolling()
{
    using CancellationTokenSource tokenSource = new();
    CancelWithPolling(tokenSource);
}
public static void CancelWithPolling
    (CancellationTokenSource tokenSource)
{
    Task.Run(() => FindSmallXValues(GeneratePoints
        (1000000), tokenSource.Token),
            tokenSource.Token);
    if (Console.ReadKey(true).KeyChar == 'x')
    {
        tokenSource.Cancel();
        Console.WriteLine("Press a key to quit");
    }
}
```

2. Next, add a new class named `MultipleTokensExample`.

3. Create a method named `CancelWithMultipleTokens` in the `MultipleTokensExample` class. This method accepts `parentToken` as a parameter, creates its own `tokenSource`, and then combines them into a `combinedSource` object to pass to the `CancelWithPolling` method:

```
public static void CancelWithMultipleTokens
    (CancellationToken parentToken)
{
    using CancellationTokenSource tokenSource = new();
    using CancellationTokenSource combinedSource =
        CancellationTokenSource.CreateLinked
            TokenSource(parentToken, tokenSource
                .Token);
    PollingExample.CancelWithPolling(combinedSource);
    Thread.Sleep(1000);
    tokenSource.Cancel();
}
```

We're calling `tokenSource.Cancel`, but if `Cancel` is invoked on any of the three `CancellationTokenSource` objects, the processing in `CancellWithPolling` will receive a cancellation request.

4. Add some code to `Program.cs` to call `CancelWithMultipleTokens`:

```
using CancellationPatterns;
CancellationTokenSource tokenSource = new();
MultipleTokensExample.CancelWithMultipleTokens
    (tokenSource.Token);
Console.ReadKey();
```

5. Run the program, and you should see an output similar to what you saw in the subsection *Canceling with polling* of the section *Discovering patterns for thread cancellation*.

Try changing which `CancellationTokenSource` object is used to invoke `Cancel`. The output should remain the same regardlesss of the source of the cancellation request.

A background `Task` will also end if you throw an exception within the `Task`. This has a similar effect of ending the background processing, but `TaskStatus` will be `Faulted` instead of `Canceled`.

This completes our review of cancellation requests from multiple sources and our tour of canceling tasks and threads with C# and .NET. Let's review what we have learned in this chapter.

Summary

In this chapter, we learned a number of new ways to cancel background threads and tasks. It is important to provide your users with a method of canceling long-running tasks or automatically canceling them when users or the operating system closes or suspends your application.

After working through the examples in this chapter, you now understand how to use polling, callbacks, and wait handles to cooperatively cancel background tasks. Additionally, you learned how to handle cancellation requests from more than one source.

In the next chapter, we will look at how .NET developers can unit test code that employs multithreaded constructs.

Questions

1. Which property of a `CancellationToken` object indicates whether a cancellation request has been made?
2. Which data type provides a `CancellationToken` object?
3. What exception type is thrown when `ThrowIfCancellationRequested` is invoked?
4. What cancellation pattern is used by the `WebClient` object in .NET?
5. Which .NET type can pause or resume operations with a `CancellationToken` object?
6. Which reset event is used to pause processing?
7. Which static method in `CancellationTokenSource` can combine multiple `CancellationToken` objects into a single source?

12
Unit Testing Async, Concurrent, and Parallel Code

Unit testing asynchronous, concurrent, and parallel code can be a challenge for .NET developers. Fortunately, there are some steps you can take to help ease the difficulty. This chapter will provide some concrete advice and useful examples of how developers can unit test code that leverages multithreaded constructs. These examples will illustrate how unit tests can still be reliable while covering code that performs multithreaded operations. In addition, we will explore a third-party tool that facilitates the creation of automated unit tests that monitor your code for potential memory leaks.

Creating unit tests for your .NET projects is important to maintain the health of your code base as it grows and evolves. When developers make changes to code that has unit test coverage, they can run the existing tests to feel confident that no existing functionality has been broken by the code changes. Visual Studio makes it simple to create, run, and maintain unit test projects throughout the life cycle of your code.

The **Test Explorer** window in Visual Studio can detect and run unit tests created with Microsoft's MSTest framework, as well as third-party frameworks such as NUnit and xUnit.net. Whether you are developing applications for Windows, mobile devices, or the cloud, you should always plan to develop a suite of unit tests for your projects and define goals for test coverage.

> **Note**
>
> This chapter assumes that you have some familiarity with unit testing and good unit testing practices. For a good primer on unit testing projects with xUnit.net, you can review Microsoft's documentation at https://docs.microsoft.com/dotnet/core/testing/unit-testing-with-dotnet-test and at https://docs.microsoft.com/visualstudio/test/getting-started-with-unit-testing.

In this chapter, we will cover the following:

- Unit testing asynchronous code

- Unit testing concurrent code

- Unit testing parallel code

- Checking for memory leaks with unit tests

By the end of this chapter, you will be armed with tools and advice to help you confidently write modern multithreaded code with unit test coverage.

> **Note**
> The unit tests in this chapter are created with the **xUnit.net** unit testing framework. You can achieve the same results with your unit testing framework of choice, including **MSTest** and **NUnit**. The memory unit testing framework we will be demonstrating later in this chapter uses xUnit.net, but it also supports MSTest and NUnit.

Technical requirements

To follow along with the examples in this chapter, the following software is recommended for Windows developers:

- Visual Studio 2022 version 17.2 or later

- .NET 6

- A JetBrains dotMemory Unit standalone console runner

All the code examples for this chapter can be found on GitHub at `https://github.com/PacktPublishing/Parallel-Programming-and-Concurrency-with-C-sharp-10-and-.NET-6/tree/main/chapter12`.

Let's get started by examining how to write unit tests that cover `async` C# methods.

Unit testing asynchronous code

Unit testing asynchronous code requires the same approach as writing good asynchronous C# code. If you need a refresher on how to work with `async` methods, you can review *Chapter 5*.

When writing a unit test for an `async` method, you will use the `await` keyword to wait for the method to complete. This requires that your unit test method is `async` and returns `Task`. Just like other C# code, creating `async void` methods is not permitted. Let's look at a very simple test method:

```csharp
[Fact]
private async Task GetBookAsync_Returns_A_Book()
{
    // Arrange
    BookService bookService = new();
    var bookId = 123;
    // Act
    var book = await bookService.GetBookAsync(bookId);
    // Assert
    Assert.NotNull(book);
    Assert.Equal(bookId, book.Id);
}
```

This probably looks like most tests you have written for synchronous code. There are only a couple of differences:

- First, the test method is `async` and returns `Task`.

- Second, the call to `GetBookAsync` uses the `await` keyword to wait for the result. Otherwise, this test follows the typical **Arrange–Act–Assert** pattern and tests the result as you typically would.

Let's create a simple project to try this in Visual Studio and see the results:

1. Start by creating a new **Class Library** project in Visual Studio named `AsyncUnitTesting`:

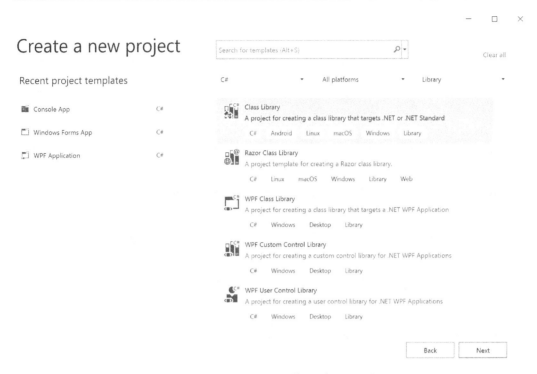

Figure 12.1 – Creating a new Class Library project

2. Next, we are going to add a test project to the **AsyncUnitTesting** solution. Right-click on the solution file in **Solution Explorer** and click on **Add | New project**. Select the **xUnit Test Project** template and name the project `AsyncUnitTesting.Tests`:

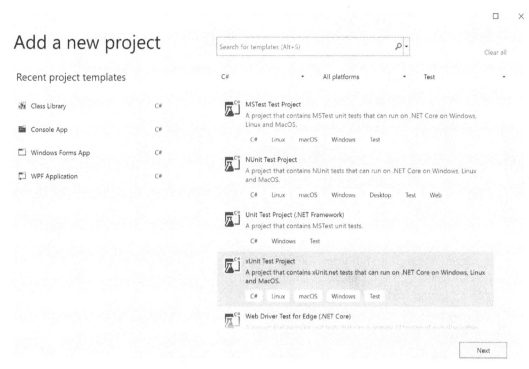

Figure 12.2 – Adding an xUnit Test project to the solution

3. In the **AsyncUnitTesting** project, rename the **Class1.cs** file `BookOrderService.cs`.
    When Visual Studio asks whether you want to rename all uses of `Class1`, select **Yes**.

4. Open the `BookOrderService` class and add an `async` method named
    `GetCustomerOrdersAsync`:

```
public async Task<List<string>>
    GetCustomerOrdersAsync(int customerId)
{
    if (customerId < 1)
    {
        throw new ArgumentException("Customer ID must
            be greater than zero.", nameof
                (customerId));
    }
    var orders = new List<string>
    {
        customerId + "1",
```

```
                customerId + "2",
                customerId + "3",
                customerId + "4",
                customerId + "5",
                customerId + "6"
    };
    // Simulate time to fetch orders
    await Task.Delay(1500);
    return orders;
}
```

This method takes `customerId` as a parameter and returns `List<string>` containing the order numbers. If `customerId` provided is less than 1, `ArgumentException` is thrown. Otherwise, a list of six order numbers is created, with `customerId` as the prefix. After injecting `Task.Delay` of `1500` milliseconds, `orders` is returned to the calling method.

5. Next, right-click the **AsyncUnitTesting.Tests** project and click on **Add | Project Reference**. In the **Reference Manager** dialog, check the box for the **AsyncUnitTesting** project and click **OK**.

6. Now, rename the `UnitTest1` class `BookOrderServiceTests` and open the file in the Visual Studio editor.

7. It's time to start adding tests. Let's start by testing the happy path. Add a test method named `GetCustomerOrdersAsync_Returns_Orders_For_Valid_CustomerId`:

```
[Fact]
public async Task GetCustomerOrdersAsync_Returns_
    Orders_For_Valid_CustomerId()
{
    var service = new BookOrderService();
    int customerId = 3;
    var orders = await service.GetCustomerOrdersAsync
        (customerId);
    Assert.NotNull(orders);
    Assert.True(orders.Any());
    Assert.StartsWith(customerId.ToString(),
        orders[0]);
}
```

After calling Get CustomerOrdersAsync with customerId of 3, our code has three assertions:

- First, we're checking that the list of orders is not null.

- Second, we're checking that the list contains some items.

- Finally, we check that the first order starts with customerId.

8. Click on **Test | Run All Tests** to ensure that this test passes.

9. Let's write that same test with a new customerId but without async and await. Assume that you have some legacy test code that just cannot be refactored, and you have to test the Get CustomerOrdersAsync method. That code would look like this:

```
[Fact]
public void GetCustomerOrdersAsync_Returns_Orders
    _For_Valid_CustomerId_Sync()
{
    var service = new BookOrderService();
    int customerId = 5;
    List<string> orders = service.GetCustomer
        OrdersAsync(customerId).GetAwaiter()
            .GetResult();
    Assert.NotNull(orders);
    Assert.True(orders.Any());
    Assert.StartsWith(customerId.ToString(),
        orders[0]);
}
```

The test method is not async and returns void. Instead of using await to allow Get CustomerOrdersAsync to run to completion, we are calling GetAwaiter(). GetResult(). The setup and assertion sections of the code remain the same.

10. Click on **Test | Run All Tests** to make sure both of our tests are *green* (passing).

11. Finally, we are going to test the exception case. Create another test but pass a negative customerId to the method under test. The entire call to Get CustomerOrdersAsync will be wrapped in an Assert.ThrowsAsync<ArgumentException> invocation:

```
[Fact]
public async Task GetCustomerOrdersAsync_
    Throws_Exception_For_Invalid_CustomerId()
{
```

```
        var service = new BookOrderService();
        await Assert.ThrowsAsync<ArgumentException>(async
            () => await service.GetCustomerOrdersAsync
                (-2));
    }
```

12. Execute **Run All Tests** one last time and ensure that they are all passing:

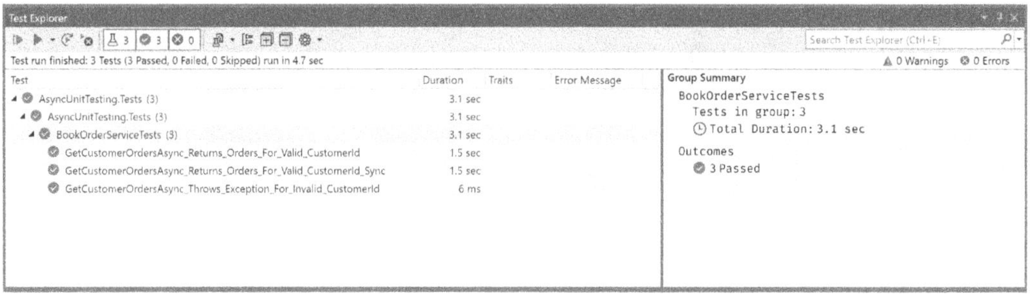

Figure 12.3 – Viewing three passing tests in Test Explorer

We now have three passing unit tests for the `GetCustomerOrdersAsync` method. The first two are essentially testing the same thing, but they are demonstrating two different ways of writing the test. You will be using the `async` method in most cases. The final test provides test coverage of the code that throws `ArgumentException`. If you are using Visual Studio Enterprise edition or a third-party tool such as dotCover, you can use their visualization tools to view which parts of your code are covered by unit tests and which are not.

Now that we have some familiarity with testing `async` methods, let's move on to working with concurrent data structures in a system under test.

Unit testing concurrent code

In this section, we will adapt a sample from *Chapter 9*, to add unit test coverage. When your code uses `async` and `await`, adding reliable test coverage is very simple. At the end of the example, we will examine an alternative method of waiting to perform your assertions by using the `SpinLock` struct.

Let's create an xUnit.net unit test project for the `ConcurrentOrderQueue` project and add several tests:

1. Start by copying the **ConcurrentOrderQueue** project from *Chapter 9*. You can get the source code from the GitHub repository if you do not already have a copy of it: `https://github.com/ PacktPublishing/Parallel-Programming-and-Concurrency-with-C- sharp-10-and-.NET-6/tree/main/chapter09/ConcurrentOrderQueue`.

2. Open the **ConcurrentOrderQueue** solution in Visual Studio.

3. Right-click the solution file in **Solution Explorer** and click on **Add | New Project**. Add an **xUnit Unit Test** project named `ConcurrentOrderQueue.Tests`. Make sure to add the new project inside the **ConcurrentOrderQueue** folder.

4. If your new test project also appears as a folder under the **ConcurrentOrderQueue** project, right-click on the **ConcurrentOrderQueue.Tests** folder and select **Exclude from Project**.

5. Add **Project Reference** from the new project to the **ConcurrentOrderQueue** project and rename the `UnitTest1` class `OrderServiceTests`.

6. In order to control which `CustomerId` values are used to generate the list of orders, we are going to create a new overload for the public `EnqueueOrders` method in the `OrderService` class:

```
public async Task EnqueueOrders(List<int> customerIds)
{
    var tasks = new List<Task>();
    foreach (int id in customerIds)
    {
        tasks.Add(EnqueueOrders(id));
    }
    await Task.WhenAll(tasks);
}
```

This method takes a list of `customerId` and calls the private `EnqueueOrders` method for each of them, adding `Task` from each call to `List<Task>` to be awaited before exiting the method.

7. We can now optimize the parameterless version of `EnqueueOrders` by having it call this new overload:

```
public async Task EnqueueOrders()
{
    await EnqueueOrders(new List<int> { 1, 2 });
}
```

8. Create a new unit test method in the `OrderServiceTests` class to test `EnqueueOrders`:

    ```
    [Fact]
    public async Task EnqueueOrders_Creates_Orders_For_
        All_Customers()
    {
        var orderService = new OrderService();
        var orderNumbers = new List<int> { 2, 5, 9 };
        await orderService.EnqueueOrders(orderNumbers);
        var orders = orderService.DequeueOrders();
        Assert.NotNull(orders);
        Assert.True(orders.Any());
        Assert.Contains(orders, o => o.CustomerId == 2);
        Assert.Contains(orders, o => o.CustomerId == 5);
        Assert.Contains(orders, o => o.CustomerId == 9);
    }
    ```

    The test will call `EnqueueOrders` with three customer IDs. After `EnqueueOrders` and `DequeueOrders` are complete, we assert that the `orders` collection is not `null`, contains some orders, and contains orders with all three of our customer IDs.

9. Run the new test and ensure that it passes.

This covers the basics of working with a system under test that uses `ConcurrentQueue`. Let's consider another scenario where we are working with code but cannot use `async` and `await` in our tests. Perhaps the method under test is not `async`. One of the tools at our disposal is the `SpinWait` struct. This struct contains some methods that provide non-locking mechanisms for waiting in our code. We will use `SpinWait.WaitUntil()` to wait until all orders have been enqueued.

The following steps will demonstrate how to reliably test the result of a method when you cannot explicitly wait for it to complete:

1. Start by adding a new public variable to the `OrderService` class to expose the number of customers whose orders have been enqueued:

    ```
    public int EnqueueCount = 0;
    ```

2. Next, increment `EnqueueCount` at the end of the private `EnqueueOrders` method:

    ```
    private async Task EnqueueOrders(int customerId)
    {
        for (int i = 1; i < 6; i++)
        {
    ```

```
        . . .
    }
    EnqueueCount++;
}
```

3. Now, create an `EnqueueOrdersSync` public method to be called from our new test. It will be similar to the public `EnqueueOrders` method. The differences between the previous example and this one are that it is not `async`, it resets `EnqueueCount` to 0, and it does not wait for the tasks to be completed:

```
public void EnqueueOrdersSync(List<int> customerIds)
{
    EnqueueCount = 0;
    var tasks = new List<Task>();
    foreach (int id in customerIds)
    {
        tasks.Add(EnqueueOrders(id));
    }
}
```

4. Next, we will create a new synchronous test method to test `EnqueueOrdersSync`:

```
[Fact]
public void EnqueueOrders_Creates_Orders_For_All
    _Customers_SpinWait()
{
    var orderService = new OrderService();
    var orderNumbers = new List<int> { 2, 5, 9 };
    orderService.EnqueueOrdersSync(orderNumbers);
    SpinWait.SpinUntil(() => orderService.EnqueueCount
        == orderNumbers.Count);
    var orders = orderService.DequeueOrders();
    Assert.NotNull(orders);
    Assert.True(orders.Any());
    Assert.Contains(orders, o => o.CustomerId == 2);
    Assert.Contains(orders, o => o.CustomerId == 5);
    Assert.Contains(orders, o => o.CustomerId == 9);
}
```

The differences are highlighted in the preceding code snippet. `SpinWait.SpinUntil` will wait without locking until the `orderService.EnqueueCount` value matches the `orderNumbers.Count`. If you want to ensure it doesn't spin forever, there are overloads for providing a timeout period as either `TimeSpan` or in milliseconds.

5.   Run the tests again and make sure that they both pass. We now have unit test methods that are testing the two methods available to enqueue orders in the `OrderService` class. In your own projects, you would add more scenarios to increase the test coverage of the class. You should always test things, such as how your code handles invalid input.

It is important to remember when unit testing multithreaded code that if you are not using `async` and `await` or some other synchronization method, your tests are going to be unreliable. Having unreliable tests is as bad as having no tests at all. Be sure to design and develop your unit tests with care. It is best to use `async/await` wherever possible for maximum reliability.

In the next section, we will build some unit tests for code that use the `Parallel.ForEach` and `Parallel.ForEachAsync` methods.

Unit testing parallel code

Creating unit tests for code that use `Parallel.Invoke`, `Parallel.For`, `Parallel.ForEach`, and `Parallel.ForEachAsync` is relatively straightforward. While they can run processes in parallel when conditions are suitable, they run synchronously relative to the invoking code. Unless you wrap `Parallel.ForEach` in a `Task.Run` statement, the flow of code will not continue until all iterations of the loop have been completed.

The one caveat to consider when testing code that uses parallel loops is the type of exceptions to expect. If an exception is thrown within the body of one of these constructs, the surrounding code must catch `AggregateException`. The exception to this `Exception` rule is `Parallel.ForEachAsync`. Because it is called with `async/await`, you must handle `Exception` instead of `AggregateException`. Let's create an example to illustrate these scenarios:

1.   Create a new **Class Library** project in Visual Studio named `ParallelExample`.

2.   Rename `Class1` `TextService` and create a method named `ProcessText` in this class:

```
public List<string> ProcessText(List<string>
    textValues)
{
    List<string> result = new();
    Parallel.ForEach(textValues, (txt) =>
    {
        if (string.IsNullOrEmpty(txt))
        {
```

```
                throw new Exception("Strings cannot be
                    empty");
            }
        result.Add(string.Concat(txt,
            Environment.TickCount));
        });
        return result;
    }
```

This method accepts a list of strings and appends `Environment.TickCount` to each value inside a `Parallel.ForEach` loop. If any of the strings are `null` or empty, `Exception` will be thrown.

3. Next, create the `async` version of `ProcessText` and name it `ProcessTextAsync`. The `async` version uses `Parallel.ForEachAsync` to perform the same operation:

```
public async Task<List<string>>
    ProcessTextAsync(List<string> textValues)
{
    List<string> result = new();
    await Parallel.ForEachAsync(textValues, async
        (txt, _) =>
    {
        if (string.IsNullOrEmpty(txt))
        {
            throw new Exception("Strings cannot
                be empty");
        }
        result.Add(string.Concat(txt,
            Environment.TickCount));
        await Task.Delay(100);
    });
    return result;
}
```

4. Add a new **xUnit Test** project to the solution and name it `ParallelExample.Tests`.

5. Rename the `UnitTest1` class `TextServiceTests` and add a **Project** reference to the **ParallelExample** project.

6. Next, we will add two unit tests to test the `ProcessText` method:

```
[Fact]
public void ProcessText_Returns_Expected_Strings()
{
    var service = new TextService();
    var fruits = new List<string> { "apple", "orange",
        "banana", "peach", "cherry" };
    var results = service.ProcessText(fruits);
    Assert.Equal(fruits.Count, results.Count);
}
[Fact]
public void ProcessText_Throws_Exception_For
    _Empty_String()
{
    var service = new TextService();
    var fruits = new List<string> { "apple", "orange",
        "banana", "peach", "" };
    Assert.Throws<AggregateException>(() =>
        service.ProcessText(fruits));
}
```

The first test calls `ProcessText` with a list of five-string values containing fruit names. The assertion checks that `results.Count` matches `fruits.Count`.

The second test makes the same call, but one of the `fruits` string values is empty. This test will ensure that `AggregateException` is thrown by the `Parallel.ForEach` loop in the method under test.

7. Add two more tests. These two tests will run the same assertions on the `ProcessTextAsync` method. The difference here is that `Assert.ThrowsAsync` must check for `Exception` instead of `AggregateExceptoin` because we are using `async/await`:

```
[Fact]
public async Task ProcessTextAsync_Returns_Expected
    _Strings()
{
    var service = new TextService();
    var fruits = new List<string> { "apple", "orange",
        "banana", "peach", "cherry" };
```

```
            var results = await service.ProcessTextAsync
                (fruits);
            Assert.Equal(fruits.Count, results.Count);
    }
    [Fact]
    public async Task ProcessTextAsync_Throws_Exception
        _For_Empty_String()
    {
        var service = new TextService();
        var fruits = new List<string> { "apple", "orange",
            "banana", "peach", "" };
        await Assert.ThrowsAsync<Exception>(async () =>
            await service.ProcessTextAsync(fruits));
    }
```

8. Run all four tests with the **Run All Tests in View** button in the **Text Explorer** window. If the window is not visible in Visual Studio, you can open it from **View | Test Explorer**. All tests should pass:

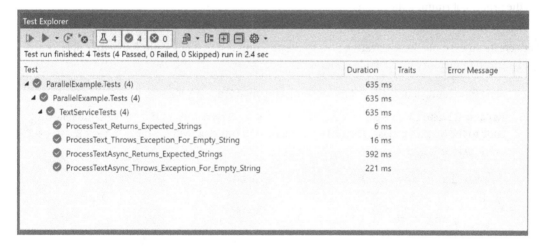

Figure 12.4 – Four tests passing in the TextServiceTests class

You now have two tests for each of the methods for processing text in the TextService class. They are testing valid and invalid input data successfully. Spend some time on your own to examine how the test coverage could be expanded. What other types of input could be used?

In the final section of this chapter, we will examine how you can build memory leak detection into your automated unit test suite.

Checking for memory leaks with unit tests

Memory leaks are by no means unique to multithreaded code, but they certainly can happen. The more code that is executing in your application, the more likely it is that some objects are going to leak. The company that makes the popular .NET tools, **ReSharper** and **Rider**, also makes a tool called **dotMemory** for analyzing memory leaks. While these tools are not free, JetBrains does offer its memory unit testing tool for free. It's called **dotMemory Unit**.

In this section, we will create a dotMemory Unit test to check whether we are leaking one of our objects. You can run these dotMemory Unit tests for free with .NET on the command line by downloading the standalone test runner here: `https://www.jetbrains.com/dotmemory/unit/`.

> **Note**
>
> For more information about using the free tooling, you can read about it here: `https://www.jetbrains.com/help/dotmemory-unit/Using_dotMemory_Unit_Standalone_Runner.html`. JetBrains also has integration for dotMemory Unit in its ReSharper and Rider tools. If you have licenses for either of these tools, it greatly simplifies the process of running these tests.

Let's create an example demonstrating how to create a unit test that determines whether objects are being leaked in memory by the code under test:

1. Start by creating a new **Class Library** project named `MemoryExample`.

2. Rename `Class1` `WorkService` and add another class named `Worker`. Add the following code to the `Worker` class. The DoWork method in this class will handle a `TimerElapsed` event in `WorkService`:

    ```
    public class Worker : IDisposable
    {
        public void Dispose()
        {
            // dispose objects here
        }
        public void DoWork(object? sender,
            System.Timers.ElapsedEventArgs e)
        {
            Parallel.For(0, 5, (x) =>
    ```

```
        {
            Thread.Sleep(100);
        });
    }
}
```

This class implements IDisposable, so we can use it with a using statement elsewhere.

3. Add a WorkWithTimer method to the WorkService class:

```
public void WorkWithTimer()
{
    using var worker = new Worker();
    var timer = new System.Timers.Timer(1000);
    timer.Elapsed += worker.DoWork;
    timer.Start();
    Thread.Sleep(5000);
}
```

This code has some problems that will prevent the worker object from being released from memory. The timer object is not stopped or disposed of, and the Elapsed event is never unhooked. When we check for leaks, we should find some.

4. Add a new **xUnit Test** project to the solution named MemoryExample.Tests.

5. Add a project reference to **MemoryExample** and add a **NuGet package reference** to **JetBrains. dotMemoryUnit**:

JetBrains.dotMemoryUnit ⊘ by JetBrains, **5.11M** downloads
dotMemory Unit is an additional unit testing framework that allows you to write tests that check code for all kinds of memory issues. For example, these can be tests that determine leaks by checking memory for objects of a particular...

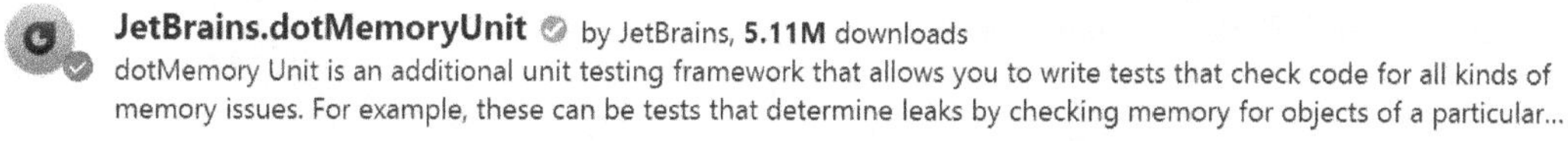

Figure 12.5 – Referencing the dotMemoryUnit NuGet package

6. Rename the UnitTest1 class in **MemoryExample.Tests** WorkServiceMemoryTests and add the following code:

```
using JetBrains.dotMemoryUnit;
[assembly: SuppressXUnitOutputExceptionAttribute]
namespace MemoryExample.Tests
{
    public class WorkServiceMemoryTests
    {
```

```
[Fact]
public void WorkWithSquares_Releases_Memory_
    From_Bitmaps()
{
    var service = new WorkService();
    service.WorkWithTimer();
    GC.Collect();
    // Make sure there are no Worker
        objects in memory
    dotMemory.Check(m => Assert.Equal(0,
        m.GetObjects(o =>
            o.Type.Is<Worker>())
                .ObjectsCount));
    }
}
}
```

A few lines are highlighted in the previous snippet. An `assembly` attribute must be added to suppress an error in the console runner when using xUnit.net with dotMemory Unit. After calling the method under test, `WorkWithTimer`, we are calling `GC.Collect` to attempt to clean all unused managed objects from memory. Finally, `dotMemory.Check` is called to determine whether there are any objects of the `Worker` type remaining in memory.

7. Run the following command either in **PowerShell** or the **Windows command line** from the folder where you downloaded and extracted the dotMemory Unit command-line tool. If you use PowerShell, the `.\` characters are required:

```
.\dotMemoryUnit.exe "c:\Program Files\dotnet\dotnet.exe"
- test "c:\dev\net6.0\MemoryExample.Tests.dll"
```

The path to .NET should be the same on your system. You will need to replace the path to `MemoryExample.Tests.dll` with your own output path where this DLL resides. The test should fail, with one `Worker` object remaining in memory, and your output will look something like this:

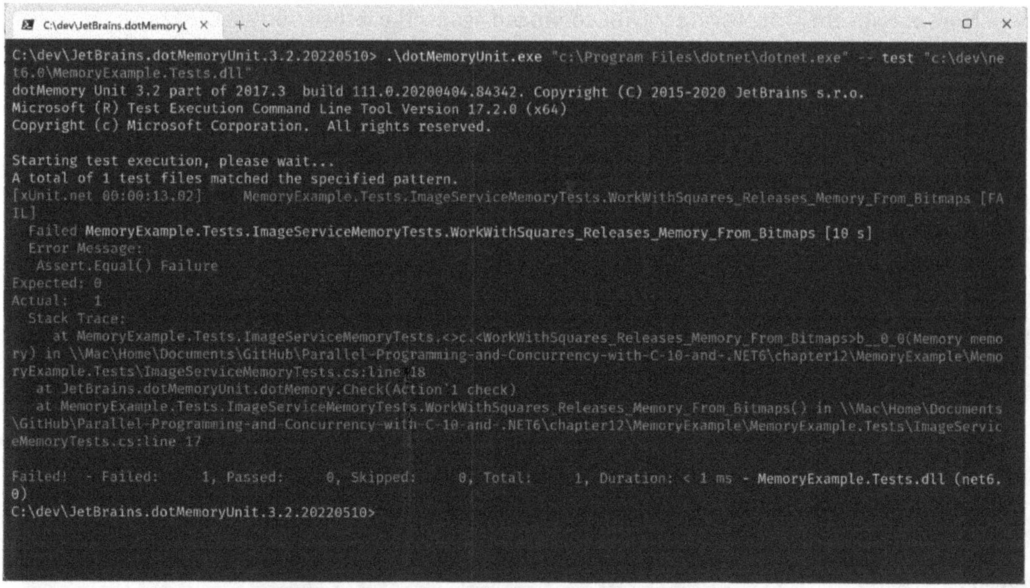

Figure 12.6 – Reviewing the failed dotMemoryUnit test run

8. In order to fix the problem, make the following changes to your `WorkService.WorkWithTimer` method:

```
public void WorkWithTimer()
{
    using var worker = new Worker();
    using var timer = new System.Timers.Timer(1000);
    timer.Elapsed += worker.DoWork;
    timer.Start();
    Thread.Sleep(5000);
    timer.Stop();
    timer.Elapsed -= worker.DoWork;
}
```

To make sure the `worker` object instance is released, we're initializing `timer` in a `using` statement, stopping `timer` when it's finished, and unhooking the `timer.Elapsed` event handler.

9. Now, execute the dotMemory Unit command again. The test should succeed now:

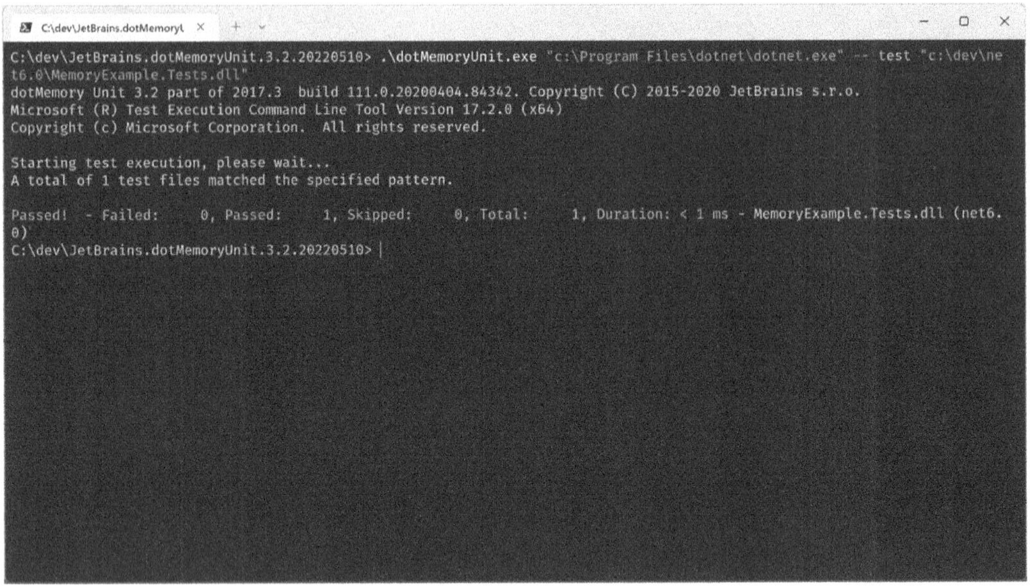

Figure 12.7 – The dotMemoryUnit test runs successfully

That concludes this example and the section on memory unit tests. If you would like to read more about dotMemory Unit, you can find its documentation here: https://www.jetbrains.com/help/dotmemory-unit/Introduction.html. The command-line tool can also be deployed to a **continuous integration** (**CI**) build server to execute these tests as part of a CI build process.

Let's finish up by reviewing what we have learned in the final chapter of this book.

Summary

In this chapter, we learned about some tools and techniques to unit test .NET projects that contain different multithreaded constructs. We started by discussing the best methods for testing C# code that employs async/await. This will be common in modern applications, and it is important to have a suite of automated unit tests covering your async code.

We also walked through some examples of unit tests that test methods that leverage parallel constructs and concurrent data structures. In the last section of the chapter, we learned about dotMemory Unit from JetBrains. This free unit testing tool adds the ability to detect objects leaked by methods under test. It is a powerful automation tool for synchronous and asynchronous .NET code.

This is the final chapter. Thanks for following along on this multithreading journey. Hopefully, you didn't encounter any deadlocks or race conditions along the way. This book provided guidance for your path through the modern, multithreaded world of .NET and C#. You should now have an understanding of the asynchronous, concurrent, and parallel methods and structures to build fast and reliable .NET applications. If you want to learn more about these topics, I suggest reading the *.NET Parallel Programming* blog (`https://devblogs.microsoft.com/pfxteam/`) and relying on the .NET documentation (`https://docs.microsoft.com/dotnet/`). You can search for documentation on any of the topics in this book to learn more.

Questions

1. What is the keyword used in .NET attributes that decorate an `xUnit.net` test method?

2. What method can you use to add `await` to your code without locks?

3. What type of exception should you expect in unit test assertions when the method under test contains a `Parallel.ForEach` loop?

4. What type of exception should you expect in unit test assertions when the method under test contains a `Parallel.ForEachAsync` loop?

5. How can you check that an object isn't `null` in an xUnit.net assertion?

6. What is the name of the window in Visual Studio where unit tests can be managed and run?

7. What are the three most popular unit test frameworks for .NET?

8. Which JetBrains products provide tooling to run dotMemory Unit tests?

Assessments

This section contains answers to questions from all chapters.

Chapter 1, Managed Threading Concepts

1. A managed thread is a thread that is created in .NET-managed code with the `System.Threading.Thread` object.

2. Set the `Thread.IsBackground` property to `true` before calling `Thread.Start()`.

3. .NET will throw a `ThreadStateException` exception.

4. .NET prioritizes managed threads mostly based on their `Thread.Priority` value.

5. `ThreadPriority.Highest`.

6. `Thread.Abort()` is not supported by .NET 6. The code will not compile.

7. Add an object parameter to the method to be started by the new thread, and pass the data when calling `Thread.Start(data)`.

8. Pass the delegate to the cancellation token's `Register` method.

Chapter 2, Evolution of Multithreaded Programming in .NET

1. `ThreadPool`

2. C# 5.0

3. .NET Framework 4.5

4. .NET Core 3.0

5. `Task`, `Task<T>`, `ValueTask`, or `ValueTask<T>`

6. `ConcurrentDictionary<TKey, TValue>`

7. `BlockingCollection<T>`

8. **Parallel LINQ (PLINQ)**

Chapter 3, Best Practices for Managed Threading

1. Singleton.

2. `ThreadStatic`.

3. A deadlock occurs when multiple threads are all waiting to access a locked resource and cannot proceed.

4. `Monitor.TryEnter`.

5. `Interlocked`.

6. `Interlocked.Add`.

7. `MaxDegreeOfParallelism`.

8. Use the `WithDegreeOfParallelism` extension method.

9. `ThreadPool.GetMinThreads()`.

Chapter 4, User Interface Responsiveness and Threading

1. `Task` or `Task<T>`.

2. `Task.WhenAll`.

3. `Task.Factory.StartNew`.

4. A background thread on `ThreadPool`.

5. `Application.Current.Dispatcher.Invoke`.

6. `this.BeginInvoke`.

7. Check the `this.InvokeRequired` property.

Chapter 5, Asynchronous Programming with C#

1. `Task.Result`.

2. `Task.WhenAll()`.

3. `Task.WaitAll()`.

4. `Task`, `Task<TResult>`, `ValueTask`, or `ValueTask<TResult>`.

5. I/O-bound operations such as a file or network access are best suited for async methods.

6. False. It is a best practice to always suffix async methods with `Async`.

7. `Task.Run`.

Chapter 6, Parallel Programming Concepts

1. `Parallel.For.`
2. `Parallel.ForEachAsync.`
3. `Parallel.Invoke.`
4. `TaskCreationOptions.AttachToParent.`
5. `TaskCreationOptions.DenyAttach.`
6. `Task.Run` will always deny child tasks from attaching. Also, `Task.Run` has no overloaded methods to provide `TaskCreationOptions`.
7. No, regular `for` and `foreach` loops can be faster if each loop iteration is fast-running and/or there are only a few iterations of the loop.

Chapter 7, Task Parallel Library (TPL) and Dataflow

The following are the answers to this chapter's questions:

1. `JoinBlock.`
2. `BufferBlock` is a propagator block.
3. `BufferBlock.`
4. `JoinTo().`
5. `Complete().`
6. `SendAsync().`
7. `ReceiveAsync().`

Chapter 8, Parallel Data Structures and Parallel LINQ

1. `AsParallel().`
2. `AsSequential().`
3. `AsOrdered().`
4. `ForAll().`
5. `AsOrdered()` can significantly decrease performance for a query.
6. `OrderBy` and `OrderByDescending`. They will default to `ParallelMergeOptions.FullyBuffered`.

7. No. PLINQ has additional overhead that can cause queries on smaller datasets or simple queries to be slower.

8. `ParallelMergeOptions.NotBuffered`.

Chapter 9, Working with Concurrent Collections in .NET

1. `BlockingCollection<T>`.

2. `ConcurrentQueue<T>`.

3. `BlockingCollection<T>`.

4. `ConcurrentDictionary<TKey, TValue>`.

5. `Enqueue()`.

6. `TryAdd()` and `TryGetValue()`.

7. No. Always add your own synchronization mechanisms when using extension methods with concurrent collections, including standard LINQ operators.

Chapter 10, Debugging Multithreaded Applications with Visual Studio

1. Use the **Attach to Process** window or set multiple startup projects in the solution file.

2. They are grouped by process.

3. Right-click in the window and select **Columns**.

4. The **Parallel Stacks** window.

5. `.PNG` files.

6. Four.

7. The **Debug Location** toolbar.

8. Click the **Flag Just My Code** button.

Chapter 11, Canceling Asynchronous Work

1. `CancellationToken.IsCancellationRequested`

2. `CancellationTokenSource`

3. `OperationCanceledException`

4. Register callback

5. `ManualResetEventSlim`

6. `ManualResetEventSlim.Reset`

7. `CancellationTokenSource.CreateLinkedTokenSource`

Chapter 12, Unit Testing Async, Concurrent, and Parallel Code

1. `Fact`

2. `SpinLock.WaitUntil`

3. `AggregateException`

4. `Exception`

5. `Assert.NotNull`

6. **Test Explorer**

7. MSTest, NUnit, and xUnit .NET

8. ReSharper, Rider, and the dotMemory Unit standalone console runner

Index

Symbols

A

V

Visual Studio
  multi-threaded debugging 220

W

wait handles
  cancellation request, handling with 254-256
WhenAll method
  using 78, 81
Windows command line 278
Windows Community Toolkit 75
Windows Forms (WinForms) 35
Windows Presentation Foundation (WPF) 74
WinForms application 119
WithMergeOptions
  using, in PLINQ queries 182
  working 183-186
work
  canceling 18
  scheduling 18

X

xUnit.net 262

Packt.com

Subscribe to our online digital library for full access to over 7,000 books and videos, as well as industry leading tools to help you plan your personal development and advance your career. For more information, please visit our website.

Why subscribe?

- Spend less time learning and more time coding with practical eBooks and Videos from over 4,000 industry professionals

- Improve your learning with Skill Plans built especially for you

- Get a free eBook or video every month

- Fully searchable for easy access to vital information

- Copy and paste, print, and bookmark content

Did you know that Packt offers eBook versions of every book published, with PDF and ePub files available? You can upgrade to the eBook version at packt.com and as a print book customer, you are entitled to a discount on the eBook copy. Get in touch with us at customercare@packtpub.com for more details.

At www.packt.com, you can also read a collection of free technical articles, sign up for a range of free newsletters, and receive exclusive discounts and offers on Packt books and eBooks.

Other Books You May Enjoy

If you enjoyed this book, you may be interested in these other books by Packt:

Enterprise Application Development with C# 10 and .NET 6 - Second Edition

Ravindra Akella, Arun Kumar Tamirisa, Suneel Kumar Kunani, Bhupesh Guptha Muthiyalu

ISBN: 9781803232973

- Design enterprise apps by making the most of the latest features of .NET 6

- Discover different layers of an app, such as the data layer, API layer, and web layer

- Explore end-to-end architecture by implementing an enterprise web app using .NET and C# 10 and deploying it on Azure

- Focus on the core concepts of web application development and implement them in .NET 6

- Integrate the new .NET 6 health and performance check APIs into your app

- Explore MAUI and build an application targeting multiple platforms - Android, iOS, and Windows

High-Performance Programming in C# and .NET

Jason Alls

ISBN: 9781800564718

- Use correct types and collections to enhance application performance
- Profile, benchmark, and identify performance issues with the codebase
- Explore how to best perform queries on LINQ to improve an application's performance
- Effectively utilize a number of CPUs and cores through asynchronous programming
- Build responsive user interfaces with WinForms, WPF, MAUI, and WinUI
- Benchmark ADO.NET, Entity Framework Core, and Dapper for data access
- Implement CQRS and event sourcing and build and deploy microservices

Packt is searching for authors like you

If you're interested in becoming an author for Packt, please visit `authors.packtpub.com` and apply today. We have worked with thousands of developers and tech professionals, just like you, to help them share their insight with the global tech community. You can make a general application, apply for a specific hot topic that we are recruiting an author for, or submit your own idea.

Share Your Thoughts

Now you've finished *Parallel Programming and Concurrency with C# 10 and .NET 6*, we'd love to hear your thoughts! Scan the QR code below to go straight to the Amazon review page for this book and share your feedback or leave a review on the site that you purchased it from.

https://packt.link/r/1803243678

Your review is important to us and the tech community and will help us make sure we're delivering excellent quality content.

www.ingramcontent.com/pod-product-compliance
Lightning Source LLC
Chambersburg PA
CBHW062106050326
40690CB00016B/3223